三峡大学来华留学英语授课品牌课程建设项目资

# MECHANICAL TESTING TECHNOLOGY

# 机械工程测试技术

主　编　陈法法
副主编　陈保家　肖文荣

江苏大学出版社
JIANGSU UNIVERSITY PRESS
镇　江

图书在版编目(CIP)数据

机械工程测试技术 ＝ Mechanical testing
technology：英文 / 陈法法主编. — 镇江：江苏大学
出版社，2022.8
ISBN 978-7-5684-1803-4

Ⅰ. ①机… Ⅱ. ①陈… Ⅲ. ①机械工程－测试技术－
高等学校－教材－英文 Ⅳ. ①TG806

中国版本图书馆 CIP 数据核字(2022)第 103408 号

机械工程测试技术
**Mechanical testing technology**

主　　编/陈法法
责任编辑/李经晶
出版发行/江苏大学出版社
地　　址/江苏省镇江市京口区学府路 301 号(邮编：212013)
电　　话/0511-84446464(传真)
网　　址/http：//press.ujs.edu.cn
排　　版/镇江市江东印刷有限责任公司
印　　刷/江苏凤凰数码印务有限公司
开　　本/787 mm×1 092 mm　1/16
印　　张/12.5
字　　数/308 千字
版　　次/2022 年 8 月第 1 版
印　　次/2022 年 8 月第 1 次印刷
书　　号/ISBN 978-7-5684-1803-4
定　　价/52.00 元

如有印装质量问题请与本社营销部联系(电话：0511-84440882)

# Preface

Whether from the perspective of advanced developed countries or China itself, the supporting and leading role of industrialization in the historical process will never change. In terms of China's domestic situation, China's manufacturing industry is still in the situation of "large in scale but not powerful enough", which is mainly due to weak independent innovation ability, poor product quality and so on. On May 8, 2015, The State Council officially issued "Made in China 2025". "Made in China 2025" emphasizes that it is necessary to follow the development trend of "Internet Plus", take the deep integration of informatization and industrialization as the main line, promote intelligent manufacturing and green manufacturing in 10 fields, and build an upgraded version of China's manufacturing industry. Among them, intelligence, green and service are the three main development directions.

Mechanical engineering testing technology plays an important role in the training of intelligent, green and service-oriented innovative talents in mechanical manufacturing industry. Mechanical engineering testing technology involves multi-disciplinary technical knowledge such as sensor technology, computer technology, signal processing technology and control technology. It is a comprehensive technology integrating electromechanical hardware and software. At present, testing and signal processing technology is developing rapidly and widely used in the field of mechanical engineering. It has become one of the theoretical foundations that mechanical students must master. Due to the mutual movement between mechanical equipment parts, most mechanical signals are dynamic signals, coupled with installation environment and its own manufacturing, mechanical testing and signal processing have their own characteristics. Therefore, according to the characteristics of testing in mechanical engineering , this book focuses on explaining the basic knowledge and its engineering application, so that readers can well master the knowledge of signal testing, analysis and processing related with mechanical engineering, and can apply the learned knowledge to solve practical problems, so as to lay a necessary foundation for further study and research.

The content of this book can be divided into two parts: basic knowledge and engineering application. The basic knowledge is developed step by step according to the basic parts involved in the test technology, such as sensors, intermediate conditioning, signal processing, recording and display instruments, etc. The engineering application mainly includes the application of stress-strain, vibration, object level and other testing technologies in mechanical engineering. In terms of content arrangement, we strive to organically combine theory and practice, and integrate more engineering practical meaning and connotation to test technical knowledge. This book is divided into 8 chapters: Chapter 1 is the introduction; Chapter 2 introduces the principle of common sensors; Chapter 3 introduces the basic characteristics of the test system; Chapter 4 introduces the signal description method; Chapter 5 introduces the signal conditioning method; Chapter 6 introduces the basis of signal analysis and processing; Chapter 7 introduces the application of mechanical testing technology; Chapter 8 introduces computer test system and virtual instrument.

This book integrates the experience and achievements of many authors in the long-term teaching and research work, and at the same time, we are deeply grateful to learn and draw the essence of teaching materials and related books at home and abroad. The mechanical engineering testing technology of Three Gorges University has been selected into the provincial excellent courses and provincial first-class undergraduate courses in Hubei Province. This textbook is also a bilingual supporting textbook for this course. Here, we would like to thank Professor Li Li of Three Gorges University, who has guided the teaching and scientific research of many young teachers, including the author of this book. In the compilation of this textbook, we also mainly refer to *Mechanical Testing Technology and Application* edited by Professor Li Li and *Mechanical Engineering Testing Technology and Application* edited by Professor Chen Baojia. We would like to express our thanks together. We would like to express our sincere thanks to Jiangsu University press and the staff, especially Wu Chun'e, who worked hard for the smooth publication of this book.

This book was jointly compiled by Chen Fafa, Chen Baojia and Xiao Wenrong, of which chapters 3, 6, 7 and 8 were compiled by Chen Fafa, chapters 2 and 4 by Chen Baojia and chapters 1 and 5 by Xiao Wenrong. Part of the text recording, modification, formatting, proofreading and other work were completed by postgraduates Cheng Mengteng, Pan Ruixue, Liu Lili, Jiang Hao, Chen Zhengkun, Chen Xueliang, Chen Xueli, etc. The final draft, review and revision of this book were completed by Chen Fafa.

Limited to the editors' knowledge level, and the fact that this course is a new course, many problems remain to be discussed, so mistakes and inadequacies in this book are inevitable. We would be deeply grateful for your comments, criticisms and suggestions.

Editors

September, 2021

# Contents

# Chapter 1

## Introduction

Testing technology is a technology that extracts the required feature information from the test signal of the tested object. With the rapid development of computer-aided design technology, MEMS technology, industrial automation technology and information technology, etc., sensors and testing technology have been more and more widely used. Whether it is engineering research, product development, or quality monitoring, performance testing, etc., these engineering applications need to use the relevant knowledge of testing technology. Testing technology is a means for human beings to understand the objective world and a basic method of scientific research. In the information age, it is particularly important to master the relevant knowledge of sensors and testing technology.

## 1.1  Main content of testing technology

Testing technology is a measurement with the property of experiment. It includes two aspects of measurement and experiment. The main purpose of testing is to obtain feature information, which is contained in certain signals that change with time or space. Therefore, the testing principles, testing methods, testing systems and signal processing methods in the application of mechanical engineering will unfold before our eyes in the subject of testing technology.

The measurement principle refers to the basic process of measurement and the theorem of physics, chemistry, and biology involved in the measurement process. For example, the piezoelectric effect is applied in the measurement of vibration acceleration with piezoelectric crystal; the electromagnetic effect is applied in the measurement of static displacement and vibration displacement with eddy current displacement sensor; the thermoelectric effect is applied in the measurement of temperature with thermocouple.

According to the specific requirements and the actual situation in the test, different test ways need to be adopted in the test process. That is the test method, including direct test and indirect test, electrical test and optical test, analog test and digital test. In the test of mechanical engineering, various mechanical indicators (generally non-electrical physical indicators) are often transformed into electrical signals for transmission, storage and processing.

The test system is composed of different kinds of test devices. Through the test system, the useful test signal can be obtained smoothly by us. It also can conduct preliminary conversion, analysis and processing for test signals by the test system.

The original signal collected by the test system often contains a lot of noise. We need to use advanced signal processing methods to convert, analyze and process the original signal, and extract the required signal features, so as to achieve our final test goal.

The tasks of testing technology mainly include the following five aspects.

① In the equipment design, through the test of new and old products, objective evaluation for product quality and performance as well as basic data for optimizing technical parameters are provided.

② In the equipment transformation, in order to tap the potential of the equipment and improve the quality, it is often necessary to measure the load, stress and process parameters of the equipment or parts, so as to provide a basis for the verification of equipment strength and bearing capacity.

③ In environmental monitoring, it is often necessary to measure the intensity and spectrum of vibration and noise, find out the vibration source through analysis, and take corresponding vibration reduction and noise prevention measures to improve working conditions and working environment.

④ The discovery of scientific laws and the birth of new formulas are inseparable from testing technology. Laws can be found from experiments to verify theoretical research results. Experiments and theories can promote each other and develop together.

⑤ In industrial automation production, the condition monitoring, quality control and fault diagnosis of equipment are realized through the test and data acquisition of process parameters.

# 1.2 The role of testing technology in mechanical engineering

Testing technology is closely related to scientific research and engineering practice. The development history of science and technology shows that many novel discoveries and breakthroughs are based on testing. Meanwhile, the development and progress of science and technology in other fields have also provided new ways and equipment for testing and promoted the development of testing technology. In the field of engineering technology, engineering research, product development, production supervision, quality control and performance test are inseparable from testing technology. In particular, the automatic control technology widely used in modern engineering technology has fused more and more test technology, and the test device has become an important part of the control system. Even in daily life appliances, such as cars, household appliances and other aspects are inseparable from testing technology. The following are some aspects of its role.

### 1.2.1    The role in mechanical vibration and structural design

In the field of industrial manufacturing, the vibration analysis of mechanical structures is an important research topic. Various mechanical vibration test signals under working conditions or artificial input excitation are obtained by various vibration sensors. Then, these signals are analyzed and processed to extract various vibration features, so as to obtain various valuable information for mechanical structure. In particular, it can analyze the causes of abnormal vibration in working machine by means of frequency spectrum analysis of mechanical vibration signal, modal analysis of mechanical structure and parameter identification technology, and then find out methods to eliminate or reduce abnormal vibration, so as to improve the performance of mechanical structure.

### 1.2.2    The role in automated production

In the industrial automation production, the process flow, product quality and equipment operation status can be monitored and controlled through the test of process parameters and data collection. As shown in Figure 1-1, in the automatic steel rolling system, the force sensor tests the rolling force of steel rolling in real time, and the thickness sensor measures the thickness of steel plate in real time. These test signals are fed back to the control system, and the control system adjusts the roll position according to the rolling force and the sheet thickness, so as to ensure the correctness of sheet rolling size and quality.

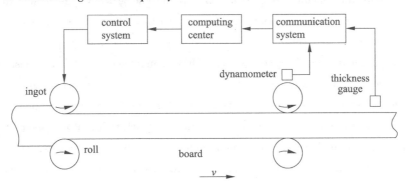

Figure 1-1    Automatic steel rolling system

### 1.2.3    The role in product quality and automatic control

At the end of the processing of automobile, machine tool equipment, motor, engine and other parts, their performance and quality must be measured and inspected. For example, in the automobile factory inspection, the measured parameters include lubricating oil temperature, cooling water temperature, fuel pressure and engine speed, etc. Through the sampling test of automobiles, engineers can understand the quality of products.

The test system is an important part in various automatic control systems. For example, various sensors play a sensory role in automatic control system. As shown in Figure 1-2, the laser ranging sensor, the robot rotation/movement position sensor and the force sensor on the welding robot of the automobile manufacturing production line work together to ensure the welding size and welding strength of the automobile body.

**Figure 1-2   Welding robot on automobile manufacturing line**

## 1.2.4   The role in mechanical monitoring and fault diagnosis

In the electric power, metallurgy, petrochemical, chemical industry and many other industries, the working conditions of some key equipment, such as steam turbine, gas turbine, water turbine, generator, motor, compressor, fan, pump and gearbox, are related to the normal production process. The implementation of 24-hour real-time dynamic monitoring for these key equipment is an essential step, which can timely and accurately provide detailed and comprehensive information for engineering and technical personnel. This is the basis of realizing the transformation from post maintenance or regular maintenance to predictive maintenance.

The stress detection of large-scale metal structure in hydropower station is shown in Figure 1-3, and the strain gauge is directly pasted on the structure for measurement. Figure 1-4 is the test system of a digital workshop, which includes cutting force sensors, processing noise sensors, ultrasonic ranging sensors, infrared proximity switch sensors, etc. These signals are transmitted to the central control room for analysis and processing as the basis for equipment monitoring and diagnosis.

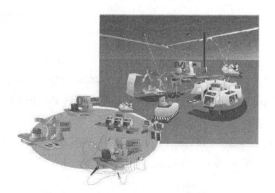

Figure 1-3   Stress detection of
metal structure

Figure 1-4   Test system of digital
workshop

In short, testing technology has been widely used in industrial and agricultural production, scientific research, domestic and foreign trade, national defense construction, transportation, medical and health, environmental protection and all aspects of people's life. It plays an increasingly important role and has become an essential basic technology for national economic development and social progress. Therefore, the use of advanced testing technology has become one of the important symbols of high economic development and scientific and technological modernization.

## 1.3   The component of test system

Information is always contained in some physical quantities and transmitted by them. These physical quantities are signals. In terms of specific physical properties, signals include electrical signals, optical signals, force signals, etc. Among them, electrical signal has obvious advantages in transformation, processing, transmission and application, so it has become the most widely used signal at present. Various non-electrical signals are often converted into electrical signals and then transmitted, processed and applied, as shown in Figure 1-5.

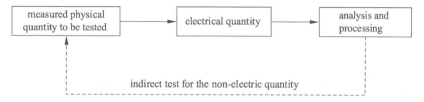

Figure 1-5   Conversion diagram from non-electric quantity to electric quantity

In many testing occasions, the specific physical properties of the signal are not considered, but are abstracted as the functional relations between variables, especially time function or space function analyzed and studied mathematically, from which some basic theories are obtained. These theories have greatly developed testing technology and become an important part of testing technology.

The basic components of the test system can be shown in Figure 1-6. Generally speaking, the test system includes sensors, signal conditioning, signal analysis, signal processing, signal display and signal recording. Sometimes the feature information that people expect to get is not directly contained in the detectable signal, then we need to find the appropriate way to stimulate the measured object to produce the signal that can fully represent its feature information and is convenient for detection.

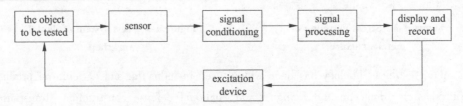

**Figure 1-6  Basic components of test system**

In the test system, when the sensor is directly affected by the measured value, it can convert the measured value into the same or different kind of output according to a certain rule, and its output is usually an electrical signal. For example, the change of mechanical strain can be converted into the change of resistance value by the metal resistance strain gauge, and the change of displacement can be converted into the change of capacitance by the capacitive sensor during the measuring of the displacement.

There are many kinds of electrical signals output by sensors, and the output power is too small. Generally, this kind of electrical signals cannot be directly put into the subsequent signal processing circuit. The main role of signal conditioning is to convert and amplify the signal, that is, to convert the signal from the sensor into a signal which is more suitable for further transmission and processing. In most cases, signal conversion is the conversion between electrical signals, converting various electrical signals into a few electrical signals which are convenient for measurement, such as voltage, current, and frequency.

In signal processing, it is to process, filter and analyze the output signal after the signal conditioning, and then output the result to display, record or control system. For example, the torque sensor can measure the speed $n$ and torque $M$ of the shaft, and the signal processing part can multiply $M$ and $n$ to get the power $P = Mn$, and then output it to the display and recording equipment.

The component of signal display and signal recording displays the measurement results in a form easy to be recognized by the observer, or stores the measurement results.

Figure 1-6 shows a complete engineering test system. In some cases, some parts can be simplified or omitted. For example, when the test system constitutes a component unit of the automatic control system, the display and recording equipment may not be needed. However, the sensors are essential for any test system.

In all these components, the basic principle must be followed is that the relationship between the output and input of each component should be kept corresponding and true as far as possible, and all kinds of interference must be reduced or eliminated as much as possible.

# 1.4　The development of testing technology

Modern testing technology is not only an important technology to promote the development of science and technology, but also the result of the development of science and technology. The development of modern science and technology constantly puts forward new requirements for testing technology and promotes the development of testing technology. At the same time, the testing technology also quickly absorb and integrate the new technologies in various scientific and technological fields such as computer, software, network, communication, etc. In recent years, the rise of new technology promotes the vigorous development of testing technology, especially in the following aspects.

## 1.4.1　Sensors are developing towards novel, miniaturized and intelligent

The role of sensors is to obtain the signal, which is the primary part in the test system. Modern test systems are all based on computers. Signal processing, conversion, storage and display are all directly related to computers, which belongs to common technology. Only sensors are ever-changing and diverse, so the role of test systems is more reflected in sensors.

Only with good and diverse sensors can these devices and technologies be effectively used in non-electric nature. The development and application of sensors are also the key to the development of superior testing devices.

The physical type sensors achieve the signal conversion by relying on the change of the physical properties of the sensitive material with the measured changes. Therefore, the development of this kind of sensor is essentially the development of new materials. The application of up-to-date knowledge, including physics, chemistry, biology, etc., to sensors is one of the important development directions of sensor technology. With the application of every new physical effect, a new type of sensitive element will appear, or a new parameter can be measured. For example, the application of new materials and components such as sound-sensitive, moisture-sensitive, color-sensitive, taste-sensitive, chemical-sensitive, and radiation-sensitive, has strongly promoted the development of sensors. Because the sensitive components of physical property sensors rely on sensitive functional materials, the development of sensitive functional materials (such as semiconductors, polymer composite materials, magnetic materials, superconducting materials, liquid crystals, biological functional materials, rare earth metals, etc.) also promotes the development of the sensor. In short, sensors are undergoing a process from structural type to physical type.

Fast changing parameters and dynamic measurement are important characteristics in mechanical engineering test and control system. Its main basis is microelectronics and computer technology. With the development of microelectronics, micro machining technology and integrated technology, many kinds of integrated sensors have appeared. The integrated sensor is an integrated arrangement of multiple sensitive elements with the same function, or an integrated body of sensitive elements with multiple different functions, or a device of an integrated body of sensors and circuits such as amplification operation and temperature compensation. The combination of sensor and microcomputer forms intelligent sensor, which is also a new trend of sensor technology development.

The intelligent sensor can automatically select the measurement range and gain, auto-calibrate and real-time calibration, carry out nonlinear correction, drift error compensation and complex calculation processing, and complete automatic fault monitoring and overload protection. By introducing advanced technology, smart sensors can use micro-processing technology to improve sensor accuracy and linearity, and correct temperature drift and time drift.

In recent years, chemical sensors have been widely used in industrial and agricultural production, environmental monitoring, medical and health care and daily life. Chemical sensors convert chemical quantities into electricity. Most chemical sensors begin to sense after the measured gas or solution molecules contact or are adsorbed by the sensitive element, and then generate corresponding current and potential. At present, the chemical sensors supplied on the market are mainly gas sensors, humidity sensors, ion sensors and biochemical sensors. It is expected that chemical sensor devices will also flourish in the future.

In recent years, sensors have also developed to many dimensions, such as manufacturing several sensors on the same substrate, and configuring the same kind of sensors into sensor arrays. Therefore, the sensor must be refined and miniaturized to form multi-dimensional sensor.

## 1.4.2 Testing instruments are developing towards high precision and multi-function

With the rapid development of computer technology and information processing technology, great changes have taken place in the test technology, which greatly improves the accuracy, measurement ability and work efficiency of the test system.

The deep combination of instrument and computer technology has formed a novel instrument structure, namely virtual instrument. The open computer architecture is adopted in virtual instrument to replace the traditional single-machine measuring instrument.

Virtual instrument integrates the common parts of traditional measuring instruments, such as power supply, operation panel, display screen, communication bus and CPU, which are shared by computers. Through the expansion board of computer instrument and application software, a variety of physical instruments are formed on the computer to achieve multi-function.

As the speed of the microprocessor increases, some real-time requirements are improved. The functions originally completed by the hardware can be realized by software, that is, the hardware functions are software-based. On the other hand, the high-speed digital processor is widely used in the test instrument, which greatly enhances the signal processing ability and performance of the instrument, and also greatly improves the accuracy of the instrument.

At the same time, by introducing many new analysis methods, the test system has the abilities of real-time analysis, memory, logic decision, self-tuning, self-adaptive control and compensation, and is developing towards multi-function.

### 1.4.3　The test and signal processing are developing towards automation

More and more test systems are using the multi-channel automatic test system with computer as the core. This system can not only realize on-line real-time measurement for dynamic parameters, but also carry out real-time signal analysis and processing quickly. With the emergence and development of signal processing unit, it has played a significant role in simplifying the structure of the signal processing system, increasing the calculation speed, and accelerating the real-time capability of signal processing.

At the same time, due to the availability of various cheap sensors and real-time processing devices, it is possible to develop multi-sensor and multi-parameter test systems. This measurement system is an essential device for automatic control system and is widely used in the field of automatic monitoring of equipment.

# 1.5　Learning objectives for this course

Testing technology involves multidisciplinary technical knowledge such as sensing technology, computer technology, signal processing technology, and control technology. It is a comprehensive technology integrating machinery and electricity, and combining hardware and software. At present, testing technology and signal processing are developing rapidly and are widely used in the field of mechanical engineering. Testing technology has become one of the theoretical bases that students of mechanical majors must master. Obviously, the testing technology has strong practicality. Therefore, students must pay attention to the theoretical learning and practical training in the study, in order to systematically master the curriculum knowledge, obtain the corresponding ability.

Students should acquire the following knowledge and abilities after completing this course.

① Master the basic theories of test technology, including common sensor principles, signal conditioning methods, basic signal analysis and processing methods, etc.

② Familiar with the testing system, testing method and computer aided testing technology for common physical quantities in mechanical engineering.

③ Have a relatively complete concept of the basic problems of dynamic testing, and can preliminarily apply the knowledge to the testing of some parameters in engineering.

# Questions

1.1   What is testing technology? What are the research objects of testing technology?

1.2   What are the basic parts of the test system? And explain the role of each part.

1.3   Try to illustrate the important role of test technology by citing the application examples of test technology around you.

1.4   Briefly summarize the development of testing technology.

1.5   How to study this course? What are the learning objectives for this course?

# Chapter 2

## Working Principle of Common Sensors

The sensor is the primary component of the test system, and it is an important device to obtain the signal of the test system. Mechanical operation status can be obtained through many types of signal detection and analysis, such as stress, vibration, noise and other signals, which need to be quantitatively described by sensors. A sensor is an output component or device that senses the signal under test directly and converts it into a corresponding physical quantity ( or signal) of another kind ( or the same kind), which is convenient for transmission and application. In mechanical testing, the sensor often converts the measured mechanical quantities ( such as force, displacement, vibration, etc.) into electrical signals that can be easily measured ( such as resistance, voltage, etc.) for display, recording, and processing using intermediate conversion circuits. The working principle, structure and characteristics of sensors commonly used in mechanical testing will be introduced in this chapter.

## 2.1　Classification of sensors

There are many types of sensors, and their working principles and applications are also different. Only by selecting the sensor correctly can we meet the various requirements of the test system and obtain the information to be measured truly and accurately. In mechanical testing, one type of indicator can be measured by multiple types of sensors. In order to facilitate the selection and application of sensors, it is necessary to classify them reasonably and scientifically. At present, the classification methods of sensors are as follows.

( 1 ) Classification by measured indicator

According to the indicator to be measured by the sensor, it can be divided into force sensor, displacement sensor, speed sensor, temperature sensor and so on. This classification method is convenient for practical selection of sensors.

(2) Classification by the physical principle of sensors

According to the physical principle of the sensor, it can be divided into mechanical sensors, electromagnetic and electronic sensors, radiation sensors, fluid sensors and so on.

(3) Classification by signal transformation characteristics of sensors

According to the transformation characteristics of sensor signals, it can be divided into two categories: physical sensors and structural sensors.

In physical type sensors, the change of the physical and chemical properties of the material of the sensitive element is corresponding to the signal to be measured. For example, the principle of mercury thermal expansion and contraction is applied in the mercury thermometer to measure the temperature.

In structural sensors, signal conversion is realized by changing the structural parameters of the sensor itself. For example, the capacitive sensor measures the capacitance by changing the distance between plates.

(4) Classification by energy conversion

According to the energy conversion of the sensor, it can be divided into energy-controlled sensors and energy conversion sensors. In energy-controlled sensors, the process of signal conversion needs the support of auxiliary power supply. Sensors such as resistive, inductive, and capacitive all belong to this category. In energy conversion sensors, the indicator can be directly measured without the need for external power supply. Such as sensors based on the piezoelectric effect.

# 2.2 Sensors based on energy control

Energy-controlled sensors are also called passive sensors. They have no inherent energy conversion during operation and cannot produce electrical signal output. They need an external power supply to work normally. This type of sensor can transform the input mechanical quantity into the change of the electrical quantity. Commonly used energy control sensors include resistive sensors, inductive sensors and capacitive sensors.

## 2.2.1 Resistive sensor

Resistive sensors are widely used to measure force, displacement, strain, torque, acceleration, etc. The basic principle is to convert the change of the measured signal into the change of the resistance value for the sensor element, and then output the voltage signal through the conversion circuit. The following description is the introduction of commonly used resistance sensors, including potentiometer sensors, resistance strain sensors and piezoresistive sensors.

## 2.2.1.1    Potentiometer sensors

(1) Structure

The structure of potentiometer sensor is shown in Figure 2-1. It consists of two basic parts: the resistance element and the brush (movable contact). The movement of the brush relative to the resistance element can be linear movement, rotation and spiral movement, so the linear displacement or angular displacement can be converted into a change in resistance or output voltage according to a certain functional relationship between them. Various potentiometer sensors can be made by using potentiometers as sensing elements. This kind of sensor can be used to measure not only linear displacement or angular displacement, but also other physical parameters which can be converted into displacement, such as pressure, acceleration and so on.

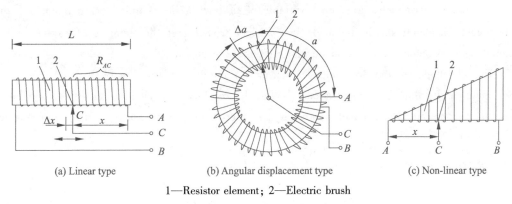

| (a) Linear type | (b) Angular displacement type | (c) Non-linear type |

1—Resistor element; 2—Electric brush

**Figure 2-1    Structure of potentiometer sensor**

(2) Working principle

Suppose the wire materials of the resistive element are the same and its cross-sectional area is constant, its resistance value changes linearly with the length of the wire. Potentiometer sensors are made based on this principle.

A linear displacement type potentiometer sensor is shown in Figure 2-1a. When the measured displacement changes, the contact $C$ will be driven to move along the potentiometer. If the moving distance is $x$, the resistance $R_{AC}$ between point $C$ and point $A$ is

$$R_{AC} = \frac{R}{L}x = K_L x \qquad (2\text{-}1)$$

Where $R$ is the total resistence of potentiometer sensor. $K_L$ is the unit resistance value of a potentiometer sensor. When the materials of the wire are the same, $K_L$ is a constant.

It can be seen that the output $R_{AC}$ of this kind of sensor has a linear relationship with the input $x$. The sensitivity of the sensor is

$$S = \frac{\mathrm{d}R_{AC}}{\mathrm{d}x} = K_L \qquad (2\text{-}2)$$

The rotary potentiometer sensor is shown in Figure 2-1b, whose resistance changes with the rotation angle, so it is called angular displacement sensors. The sensitivity of the sensor can be expressed as

$$S = \frac{\mathrm{d}R_{AC}}{\mathrm{d}\alpha} = K_\alpha \qquad (2\text{-}3)$$

Where $K_\alpha$ is the resistance value corresponding to the unit radian. When the material of the wire is uniform, $K_\alpha$ is a constant; $\alpha$ is the rotation angle.

The non-linear potentiometer sensor is shown in Figure 2-1c. The relationship between output resistance and sliding contact displacement is a nonlinear function. Various mapping relations (including exponential function, trigonometric function, logarithmic function and other arbitrary functions) between input and output can be realized in sensors with this kind of structure. The skeleton shape of the potentiometer is determined by the required output.

For example, suppose the input is $f(x) = kx^2$, where $x$ is the input displacement, in order to make the output variable $R_x$ and input variable $f(x)$ have a linear functional relationship, the potentiometer skeleton should be made into a right triangle; if the input is $f(x) = kx^3$, the potentiometer skeleton should be made into a parabolic skeleton.

The advantages of the potentiometer sensor are simple structure, stable performance, and convenient application. The disadvantage is that the resolution is not high and there is a lot of noise in the output. Potentiometer sensors are often used for measuring linear and angular displacement.

### 2.2.1.2 Resistance strain sensors

Resistance strain sensors can be used to measure strain, force, displacement, torque and other parameters. It has the advantages of small size, fast dynamic response, high accuracy and easy to use. Therefore, it has been widely used in aviation, shipbuilding, machinery, construction and other industries.

The core element of resistance strain sensor is resistance strain gauge. When the test piece or the elastic sensitive element is driven by the measured object, then the displacement, stress and strain will appear in this process. The resistance strain gauge attached to the test piece or the elastic sensitive element will convert the strain (displacement, stress) into a change in resistance. In this way, the magnitude of the measured variable can be quantified by measuring the change in resistance value of the resistance strain gauge.

(1) Structure and classification

The structure of the resistance strain gauge is shown in Figure 2-2. The resistance strain gauge is made of a high-resistance wire with a diameter of 0.025 mm. In order to obtain high resistance, the resistance wires are arranged in a grid shape, which is called sensitive grid, and adhered to the insulating substrate. A protective covering layer is pasted on the sensitive grid. The two ends of the resistance wire are welded with

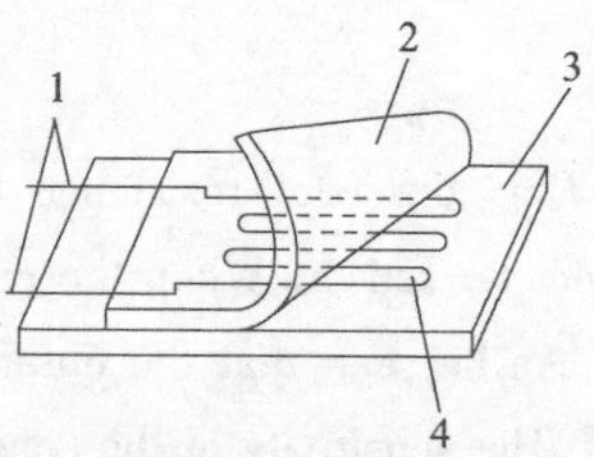

1— Lead; 2—Cladding; 3—Substrate;
4—Resistance wire

**Figure 2-2  Basic structure of resistance strain gauge**

leads.

According to the different materials and manufacturing process of the sensitive grid of the resistance strain gauge, its structural forms include wire type, foil type and membrane type, as shown in Figure 2-3.

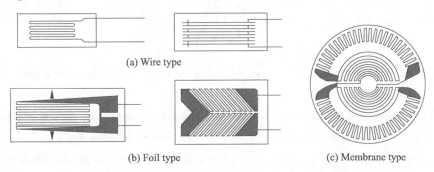

(a) Wire type

(b) Foil type                    (c) Membrane type

**Figure 2-3   Different structures of resistance strain gauge**

(2) Working principle

When a metal conductor is subjected to mechanical deformation (elongation or shortening) under the action of external forces, its resistance value will change with the deformation, which is called the resistance strain effect of the metal.

Take the wire strain gauge as an example, if the length of the wire is $l$, the cross-sectional area is $A$, the resistivity is $\rho$ and its resistance without force is $R$. According to Ohm's Law, the following formula is valid

$$R = \rho \frac{l}{A} \tag{2-4}$$

When the metal wire is deformed by the external force, all parameters including length $l$, cross-sectional area $A$ and resistivity $\rho$ will change more or less, resulting in the change of wire resistance $R$. When each parameter is changed in increments $dl$, $dA$, and $d\rho$, the resulting resistance increment $dR$ is

$$dR = \frac{\partial R}{\partial l}dl + \frac{\partial R}{\partial A}dA + \frac{\partial R}{\partial \rho}d\rho \tag{2-5}$$

Where $A = \pi r^2$, $r$ is the wire radius.

Then, the above equation can be transformed into the following form

$$\frac{dR}{R} = \frac{dl}{l} - 2\frac{dr}{r} + \frac{d\rho}{\rho} \tag{2-6}$$

In this equation, $\frac{dl}{l} = \varepsilon$ is the axial strain of the wire; $\frac{dr}{r}$ is the transverse strain of the wire diameter.

It can be known from the knowledge of material mechanics

$$\frac{dr}{r} = -\mu \frac{dl}{l} = -\mu\varepsilon \tag{2-7}$$

In the formula, $\mu$ is the Poisson coefficient of the wire material.

According to the comprehensive calculation of Equation (2-7) and Equation (2-6), the following equation can be obtained

$$\frac{\mathrm{d}R}{R} = (1+2\mu)\varepsilon + \frac{\mathrm{d}\rho}{\rho} \qquad (2\text{-}8)$$

Set

$$S_0 = \frac{\mathrm{d}R/R}{\varepsilon} = (1+2\mu) + \frac{\mathrm{d}\rho/\rho}{\varepsilon} \qquad (2\text{-}9)$$

Where $S_0$ is the sensitivity of the metal wire, and its physical meaning is the relative change in resistance caused by unit strain.

It can be seen from equation (2-9) that the sensitivity of metal materials is affected by two factors: one is caused by the geometric size change of the material after being stressed, which is $1+2\mu$; the other is caused by the change of material resistivity after the stress, that is $(\mathrm{d}\rho/\rho)/\varepsilon$. For metal materials, $(\mathrm{d}\rho/\rho)/\varepsilon$ is much smaller than $1+2\mu$.

A large number of experiments have shown that within the tensile limit of the wire, the relative change of resistance is proportional to the axial strain it receives, that is, $S_0$ is a constant. So Equation (2-8) can be simplified to the following form

$$\frac{\mathrm{d}R}{R} = S_0\varepsilon \qquad (2\text{-}10)$$

Generally, the sensitivity $S_0$ of the resistance wire is between 1.7 and 3.6.

## 2.2.1.3  Piezoresistive sensors

Metal wire and foil resistance strain gauges have stable performance and high accuracy. They are still being continuously improved and developed, and have been widely used in some high-precision strain gauge sensors. However, the main disadvantage of this type of sensor is the sensitivity is small. In order to improve this shortcoming, semiconductor strain gauge and diffusion type semiconductor strain gauge were invented at the end of 1950s. Sensors made of semiconductor strain gauges are called solid-state piezoresistive sensors. Its outstanding advantages are high sensitivity, small size, small lateral effects, small hysteresis and creep. Therefore, this type of sensor is suitable for dynamic measurement, but its main disadvantage is poor temperature stability and needs to be used under temperature compensation or constant temperature conditions.

(1) Working principle

When the semiconductor material is subjected to stress, its resistivity will change significantly. This phenomenon is called the piezoresistive effect. In fact, any material presents piezoresistive effect in varying levels, but this effect in the semiconductor materials is particularly strong. The analysis formula of resistance strain effect is applicable to semiconductor resistance materials, so formula (2-8) can still be used to describe the piezoresistive effect of semiconductors. For the metal material, $\mathrm{d}\rho/\rho$ is relatively small, but for semiconductor materials, $\mathrm{d}\rho/\rho \gg (1+2\mu)\varepsilon$, that is, the resistance change caused by mechanical deformation can be ignored, and the change rate of resistance is mainly caused by $\mathrm{d}\rho/\rho$, that is:

$$dR/R = (1+2\mu)\,\varepsilon + d\rho/\rho \approx d\rho/\rho \qquad (2\text{-}11)$$

Set

$$d\rho/\rho = \pi_L \sigma = \pi_L E \varepsilon \qquad (2\text{-}12)$$

Where $\pi_L$ is the piezoresistance coefficient along a certain crystal direction $L$; $\sigma$ is the stress along a crystal direction $L$; $E$ is the elastic modulus of the semiconductor material. Therefore, the sensitivity $S_0$ of the semiconductor material is

$$S_0 = \frac{dR/R}{\varepsilon} = \pi_L E \qquad (2\text{-}13)$$

For semiconductor silicon, $\pi_L = (40\sim80)\times10^{-11}\ \mathrm{m^2/N}$, $E = 1.67\times10^{11}\ \mathrm{Pa}$, then $S_0 = \pi_L E = 50\sim100$. Obviously, the sensitivity of semiconductor resistance material is much higher than that of metal wire.

(2) Application

Based on the characteristics of semiconductor materials, piezoresistive sensors can be divided into semiconductor strain sensors and solid piezoresistive sensors. The structure of semiconductor strain gauge sensor is shown in Figure 2-4, and its utilization is similar to that of resistance strain gauge. Solid state piezoresistive sensors are mainly used for the measurement of pressure and acceleration. The solid-state piezoresistive pressure sensor is shown in Figure 2-5. Its core part is a circular N-type silicon diaphragm which is fixed and supported around, on which there are four P-type resistors with equal resistance. The periphery of the silicon diaphragm is fixed with a silicon cup, and there are two pressure chambers on both sides, one is the high pressure chamber connected with the measured pressure, the other is the low pressure chamber connected with the atmosphere. Under the action of the measured pressure, the silicon diaphragm produces stress and strain, and the pressure is measured by the piezoresistive effect produced by the P-type resistance.

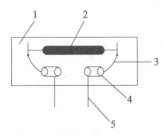

1—Substrate; 2—Semiconductor material sheet; 3—Inner lead; 4—Welding plate; 5—Outer lead

**Figure 2-4   Structure of semiconductor strain gauge**

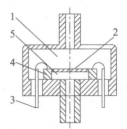

1—Low pressure cavity; 2—N-type silicon diaphragm; 3—Lead; 4—Silicon cup; 5—High pressure cavity

**Figure 2-5   Structure of solid-state piezoresistive pressure sensor**

## 2.2.2   Inductive sensor

The inductive sensor is based on the principle of electromagnetic induction, which converts the measured non-electric quantity (such as displacement, pressure, vibration, etc.) into the change of inductance. According to different conversion means, it can be divided into self-inductance sensor (including variable reluctance type and eddy current type) and mutual inductance sensor (i.e. differential transformer type).

### 2.2.2.1 Variable magnetic-resistance type sensors

The structure of the variable magnetic-resistance type sensor is shown in Figure 2-6. It consists of coil, iron core and armature. The iron core and the armature are made of magnetic material, and there is an air gap between the iron core and the armature. The thickness of the air gap is $\delta$, and the moving part of the sensor is connected with the armature. According to the knowledge of electrical engineering, the self-inductance of the coil is

$$L = \frac{W^2}{R_m} \tag{2-14}$$

In the formula, $W$ is the number of coil turns; $R_m$ is the total magnetic resistance of the magnetic circuit.

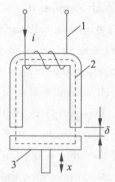

1— Coil; 2— Iron core; 3— Armature

**Figure 2-6   Structure of a variable reluctance inductive sensor**

If the iron loss of the magnetic circuit is ignored, and the air gap $\delta$ is small, the total magnetic resistance of the magnetic circuit is

$$R_m = \frac{l}{\mu A} + \frac{2\delta}{\mu_0 A_0} \tag{2-15}$$

In which, $l$ is the magnetic conductivity length of the iron core; $\mu$ is the magnetic permeability of the iron core; $A$ is the magnetically permeable cross-sectional area of the iron core; $\delta$ is the thickness of the air gap; $\mu_0$ is the air permeability; $A_0$ is the magnetically permeable cross-sectional area of the air gap.

Because $\mu \gg \mu_0$

$$R_m \approx \frac{2\delta}{\mu_0 A_0} \tag{2-16}$$

Self-induction $L$ can be written as

$$L = \frac{W^2 \mu_0 A_0}{2\delta} \tag{2-17}$$

Equation (2-17) shows that when the number of coil turns $W$ is constant, as long as $\delta$ or $A_0$

is changed, the inductance can be changed. As long as the change of inductance can be measured, the change of measured object can be determined. Therefore, the variable magnetic-resistance sensors can be divided into variable air gap type and variable air gap magnetically permeable area type.

When the cross-sectional area of air gap magnetic conductivity $A_0$ is kept constant and the thickness of air gap $\delta$ is changed, a variable air gap type sensor can be constructed. The relationship between $L$ and $\delta$ is non-linear, and its output characteristic curve is shown in Figure 2-7. The sensitivity of the sensor is

$$S=\frac{\mathrm{d}L}{\mathrm{d}\delta}=-\frac{W^2\mu_0 A_0}{2\delta^2} \tag{2-18}$$

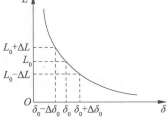

Figure 2-7　Output characteristics of variable air gap inductive sensor

Similarly, when the air gap thickness $\delta$ is kept constant and the air gap magnetic permeability cross-sectional area $A_0$ is changed, a variable air gap magnetically permeable area type sensor can be constructed. The linear relationship between self-inductance $L$ and $A_0$ is shown in Figure 2-8.

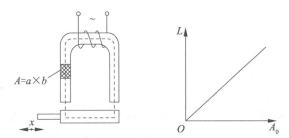

Figure 2-8　Variable air gap magnetically permeable area type sensor

The typical structures of commonly used variable magnetic-resistance sensors are shown in Figure 2-9. The variable magnetically permeable area type sensor is shown in Figure 2-9a, the self-inductance $L$ and $A_0$ have a linear relationship, and this kind of sensor has low sensitivity. Figure 2-9b is of differential type magnetic-resistance sensor. When the armature starts to move slightly, the gap between the two coils can be changed between $\delta_0+\Delta\delta$ and $\delta_0-\Delta\delta$. The self-induction of one coil increases gradually, and that of the other coil decreases. A single screw coil type is shown in Figure 2-9c. When the core moves in the coil, the magnetic resistance will change, and then the self-induction of the coil will change. This kind of sensor is simple in structure, easy to manufacture, but low in sensitivity, and suitable for large displacement measurement. A double solenoid coil differential type is shown in Figure 2-9d. It has higher sensitivity than single solenoid coils and is often used in inductive micrometers.

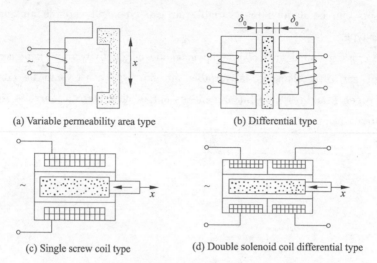

(a) Variable permeability area type      (b) Differential type

(c) Single screw coil type      (d) Double solenoid coil differential type

**Figure 2-9　Typical structures of variable reluctance sensor**

## 2.2.2.2　Eddy current sensors

According to the principle of electromagnetic induction, when a metal conductor is placed in a changing magnetic field or cuts the magnetic force line in the magnetic field, a closed eddy current will generate in the conductor, which is called eddy current effect. The sensor based on eddy current effect is called eddy current sensor. According to the penetration form of eddy current in a metal conductor, eddy current sensors are often divided into two types, high-frequency reflection type and low-frequency transmission type, but the working principles of them are basically similar.

（1）High-frequency reflective eddy current sensors

The working principle of high-frequency reflective eddy current sensor is shown in Figure 2-10. The metal plate is placed close to a coil with a spacing of $\delta$. When a high frequency alternating current $i$ is passed through the coil, the high frequency electromagnetic field generated acts on the surface of the metal plate. Eddy current $i_1$ is generated in the thin layer on the surface of the metal plate, which in turn generates a new alternating magnetic field. According to Lenz's Law, the alternating electromagnetic field of the eddy current will resist the change of the magnetic field of the coil, causing the equivalent impedance $Z$ of the original coil to change, and the change level is related to the distance $\delta$. The analysis shows that besides the distance $\delta$ between the coil and the metal plate, the resistivity $\rho$ of the metal plate, the permeability $\mu$ and the exciting circle frequency $\omega$ of the coil are also the factors that affect the impedance $Z$ of the high frequency coil. When one of these factors is changed, different mechanical quantities can be measured. For example, if $\delta$ is changed, it can be used as displacement and vibration measurement; if $\rho$ or $\mu$ is changed, it can be used as material identification or flaw detection.

(2) Low-frequency transmission eddy current sensors

The working principle of low-frequency transmission eddy current sensor is shown in Figure 2-11. The transmitting coil $L_1$ and the receiving coil $L_2$ are respectively placed above and below the metal plate to be tested. When a low-frequency voltage $u_s$ is connected to both ends of the coil $L_1$, the coil $L_1$ will generate an alternating magnetic field $\Phi_1$. If there is no metal plate between the two coils, the alternating magnetic field will cause the coil $L_2$ to generate an induced voltage $u_2$. If the metal plate under test is placed between two coils, the magnetic field generated by coil $L_1$ will generate eddy currents in the metal plate. Just then, the magnetic field energy is lost, and the magnetic field reaching to $L_2$ will be weakened to $\Phi_1'$, so that the induced voltage $u_2$ generated by $L_2$ will drop. Experiments and theories have proved that the thicker the metal plate, the greater the eddy current loss and the smaller the voltage $u_2$. Therefore, the thickness of the metal plate can be obtained according to the voltage $u_2$. The thickness detection range of the transmission eddy current sensor is $1 \sim 100$ mm, and the resolution is 0.1 μm.

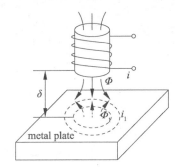

**Figure 2-10　High-frequency reflective eddy current type sensor**　　**Figure 2-11　Low-frequency transmission eddy current type sensor**

(3) Application

The most important characteristic of eddy current sensors is that it can realize non-contact continuous measurement for displacement, thickness, surface temperature, speed, stress, material damage, etc. In addition, it also has the advantages of small size, high sensitivity and wide frequency response. Therefore, in recent years, eddy current displacement and vibration measuring instruments, thickness gauges and non-destructive flaw detectors have been widely used in machinery and metallurgy. In fact, this kind of sensors can be applied for measuring radial vibration, rotary shaft error motion, rotational speed and surface crack and defect, as shown in Figure 2-12.

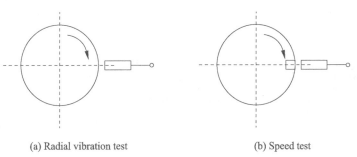

(a) Radial vibration test　　　　　　　　　(b) Speed test

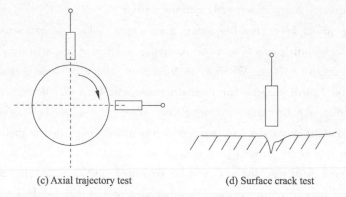

(c) Axial trajectory test            (d) Surface crack test

**Figure 2-12　Engineering application example of eddy current sensor**

### 2.2.2.3　Differential transformer sensors

（1）Mutual induction

In electromagnetic induction, the mutual induction is very common, as shown in Figure 2-13. When AC current $i_1$ is put into the coil $L_1$, the coil $L_2$ generates induced electromotive force $e_{12}$, whose magnitude is proportional to the change rate of current $i_1$, that is

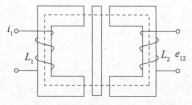

$$e_{12} = -M \frac{\mathrm{d}i_1}{\mathrm{d}t} \qquad (2\text{-}19)$$

**Figure 2-13　Mutual inductance phenomenon**

In which, $M$ is a proportional coefficient, which is called mutual inductance, and the value is related to factors such as the relative position of the coils and the magnetic conductivity of the surrounding medium.

（2）Structure and working principle

Based on this principle, the differential transformer sensor converts the measured displacement into the change of coil mutual inductance. The structure and working principle of the differential transformer sensor are shown in Figure 2-14a and b. It is mainly composed of coil, iron core and movable armature. Essentially, the structure of the coil is similar to that of a transformer, consisting of a primary coil and two secondary coils. When the primary coil is connected to a stable AC power source, the two secondary coils generate induced electromotive forces $e_1$ and $e_2$. In practice, two secondary coils are often used to form a differential type. Therefore, the output voltage of the sensor is the difference between $e_1$ and $e_2$, that is $e_0 = e_1 - e_2$, and the value is related to the position of the movable armature. When the movable armature is in the center position, $e_1 = e_2$, the output voltage $e_0 = 0$; when the active armature moves up, that is, $e_1 > e_2$, $e_0 > 0$; when moving down, $e_1 < e_2, e_0 < 0$. As the moving armature moves away from the center position, the output voltage will gradually increase, and its output characteristic is shown in Figure 2-14c.

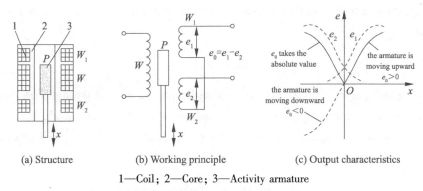

(a) Structure          (b) Working principle          (c) Output characteristics

1—Coil; 2—Core; 3—Activity armature

**Figure 2-14  Differential transformer type sensor**

The differential transformer type sensor has the advantages of high accuracy (the order of magnitude can reach 0.1 μm), good stability, easy to use, etc., and is mostly used for the measurement of linear displacement. With the help of elastic elements, physical quantities such as pressure and weight can be converted into displacement changes, so this kind of sensor can also be used to measure physical quantities such as pressure and weight.

## 2.2.3  Capacitive sensor

A capacitive sensor is a sensor that converts the change of measured mechanical quantity into the change of electrical capacity. It has the advantages of simple structure, small size, high resolution, non-contact measurement, and can work in high temperature, radiation, strong vibration and other harsh conditions. It is often used in the measurement of pressure, liquid level, vibration, displacement and other physical quantities.

The basic principle of capacitive sensors is based on the relationship between capacitance and its structural parameters.

As shown in Figure 2-15, taking the simple plate capacitor as an example, it can be derived from the physics knowledge, when the influence of edge electric field is ignored, the capacitance $C$ is

$$C = \frac{\varepsilon A}{\delta} \quad\quad (2\text{-}20)$$

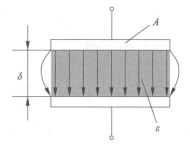

**Figure 2-15  Parallel plate capacitor**

In the formula, $\varepsilon$ is the dielectric constant, $A$ is the area of the plates, and $\delta$ is the distance between two parallel plates.

Equation (2-20) shows that when the change of measured mechanical quantity leads to the change of $\delta$, $A$ or $\varepsilon$, it will cause the capacitance $C$ to change. If two of the parameters are kept constant and only another parameter is changed, the change in that parameter can be converted into the change in capacitance.

According to the changing parameters of the capacitor, it can be divided into three types of capacitor sensors: variable electrode distance type, variable area type and variable dielectric constant type. In practice, variable electrode distance and variable area capacitive sensors are widely used.

### 2.2.3.1  Variable distance capacitive sensors

According to Equation (2-20), if the area of the two plates and the dielectric between the two plates remain unchanged, the capacitance $C$ and the electrode distance $\delta$ will have a nonlinear relationship, as shown in Figure 2-16. When the electrode distance has a small change $d\delta$, the capacitance change caused is $dC$

$$dC = -\frac{\varepsilon A}{\delta^2}d\delta \tag{2-21}$$

Thus, the sensitivity of the sensor can be obtained as

$$S = \frac{dC}{d\delta} = -\frac{\varepsilon A}{\delta^2} = -\frac{C}{\delta} \tag{2-22}$$

It can be seen that the sensitivity $S$ is inversely proportional to the square of the polar distance, and the smaller the polar distance, the higher the sensitivity. Obviously, the capacitance $C$ has a non-linear relationship with the polar distance $\delta$, so non-linear errors will be caused inevitably. In order to reduce errors, it is usually stipulated to work within a small range of the polar distance. It is generally agreed that the range of polar distance change is $\Delta\delta/\delta \approx 0.1$. At this time, the sensitivity of the sensor is approximately constant, and the output $C$ and $\delta$ have an approximately linear relationship. In practical application, in order to improve the sensitivity and working range of the sensor and overcome the influence of external conditions (such as power supply voltage, ambient temperature, etc.) on the measurement accuracy, a differential capacitance sensor is often used.

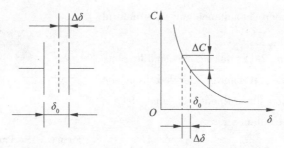

**Figure 2-16　Variable distance capacitive sensor**

### 2.2.3.2  Variable area capacitive sensors

The working principle of the variable area capacitive sensor is to change the effective area of the electrode plate under the action of the measured mechanical quantity. There are two kinds of commonly used types: angular displacement type and linear displacement type, as shown in Figure 2-17.

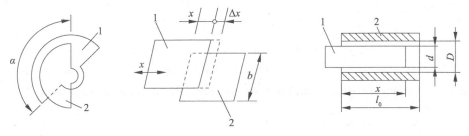

(a) Angular displacement type　　(b) Linear displacement type　　(c) Cylinder moves with linear displacement

1—Moving plate; 2—Fixed plate

**Figure 2-17　Variable area capacitive sensor**

The angular displacement capacitive sensor is shown in Figure 2-17a. When the moving plate rotates at a certain angle, the mutual coverage area between the moving plate and the fixed plate will change, resulting in changes in capacitance. The coverage area $A$ is

$$A = \frac{\alpha r^2}{2} \tag{2-23}$$

In which, $\alpha$ is the center angle corresponding to the coverage area; $r$ is the radius of the plate. The capacitance is

$$C = \frac{\varepsilon \alpha r^2}{2\delta} \tag{2-24}$$

The sensitivity is

$$S = \frac{dC}{d\alpha} = \frac{\varepsilon r^2}{2\delta} = \text{constant} \tag{2-25}$$

The output is linear with the input.

The planar linear displacement sensor is shown in Figure 2-17b. When the moving plate moves along the direction of $x$, the coverage area changes and the capacitance also changes accordingly. The capacitance $C$ is

$$C = \frac{\varepsilon b x}{\delta} \tag{2-26}$$

In the formula, $b$ is the width of the plate.

The sensitivity is

$$S = \frac{dC}{dx} = \frac{\varepsilon b}{\delta} = \text{constant} \tag{2-27}$$

The cylindrical linear displacement capacitive sensor is shown in Figure 2-17c. The moving plate (inner cylinder) and the fixed plate (outer cylinder) cover each other. When the covering length is $x$, the capacitance is

$$C = \frac{2\pi \varepsilon x}{\ln(D/d)} \tag{2-28}$$

In which, $D$ is the diameter of the outer cylinder; $d$ is the diameter of the inner cylinder.

The sensitivity is

$$S = \frac{\mathrm{d}C}{\mathrm{d}x} = \frac{2\pi\varepsilon}{\ln(D/\mathrm{d})} = \text{constant} \tag{2-29}$$

The advantage of the variable area capacitance sensor is that the output and input have a linear relationship, but the sensitivity is lower than that of the variable polar distance type, which is suitable for the measurement of large angular displacements or linear displacements.

### 2.2.3.3 Variable dielectric capacitive sensors

This kind of sensor can convert the measured mechanical quantity into a change in electricity by using the change of dielectric parameters. It can be used to measure the thickness of the dielectric (Figure 2-18a), displacement (Figure 2-18b) and liquid level (Figure 2-18c). It can also be used to measure temperature, humidity according to the changes of the dielectric constant with temperature, humidity, etc. (Figure 2-18d).

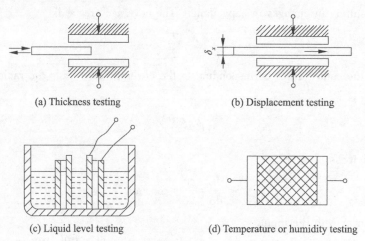

(a) Thickness testing  (b) Displacement testing

(c) Liquid level testing  (d) Temperature or humidity testing

**Figure 2-18  Variable dielectric capacitive sensor**

### 2.2.3.4 Applications of capacitive sensors

The working principle of a capacitive speed sensor is shown in Figure 2-19. In this figure, the outer edge of the gear is the moving polar plate of the capacitance sensor. When the fixed polar plate of the capacitor is opposite to the tooth top, the capacitance is the largest, and when it is opposed to the tooth gap, the capacitance is the smallest. When the gear rotates continuously, the capacitance changes periodically. The change of capacitance is converted into pulse signals through the measuring circuit, and the frequency displayed by the measured instrument represents the speed of the gear. If the number of teeth is $z$ and the frequency is $f$, the speed $n$ is

$$n = \frac{60f}{z}$$

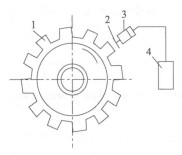

1—Gear; 2 —Fixed polar plate; 3—Capacitive sensor; 4—Frequency meter

**Figure 2-19  Working principle of the capacitive speed sensor**

The working principle of the capacitive thickness gauge for measuring the thickness of metal strips during rolling is shown in Figure 2-20. Two capacitors $C_1$ and $C_2$ are formed between the working plate and the metal strip, and the total capacitance is the sum of the two capacitors, that is $C = C_1 + C_2$. When the thickness of metal strip changes during rolling, the capacitance will be changed. Through the detection circuit, the thickness of the metal strip can be converted and displayed in the measured instrument.

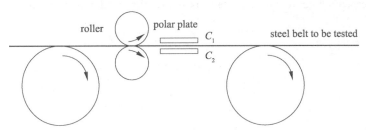

**Figure 2-20  Working principle of capacitive thickness gauge**

# 2.3  Sensors based on energy conversion

Energy conversion type sensors are also called active sensors. They have inherent energy conversion during operation and can generate electrical signal output without the need for an external power supply. This type of sensor can directly convert the input mechanical quantity into electrical signal output. Commonly used energy conversion sensors include piezoelectric sensors, magnetoelectric sensors, photoelectric sensors, Hall sensors, and pyroelectric sensors.

## 2.3.1  Piezoelectric sensor

Piezoelectric sensors are reversible converters that can convert mechanical energy into electrical energy as well as electrical energy into mechanical energy. Its working principle is based on the piezoelectric effect of certain substances. Piezoelectric sensors are small in size, light in weight, simple in structure, reliable in operation, and suitable for dynamic measurement. At present, piezoelectric sensors are mostly used for acceleration and dynamic force or pressure measurement.

### 2.3.1.1 Piezoelectric effect and piezoelectric materials

When some substances are forced in a certain direction to deform, they will generate charges on certain surfaces. When the external force is removed, they will return to the state of no charge again. This phenomenon is called piezoelectric effect. On the contrary, if an electric field is applied in the polarization direction of these substances, mechanical deformation or mechanical stress will occur in a certain direction. When the external electric field is removed, these deformations or stress also disappear. This phenomenon is called inverse piezoelectric effect.

Materials with piezoelectric effect are called piezoelectric materials, and quartz is a kind of commonly used piezoelectric material. The crystal shape of the quartz crystal is a hexagonal crystal column, as shown in Figure 2-21a. The two ends are a symmetrical pyramid, and the hexagonal prism is its basic structure. The vertical axis $z-z$ is called the optical axis, the axis $x-x$ perpendicular to the optical axis through the hexagonal ridge line is called the electrical axis, and the axis $y-y$ perpendicular to the edge face is called the mechanical axis, as shown in Figure 2-21b.

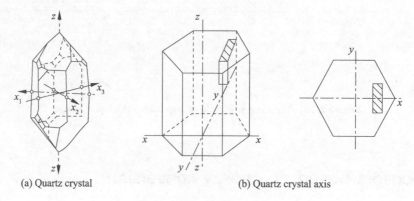

| (a) Quartz crystal | (b) Quartz crystal axis |

**Figure 2-21  Quartz crystal**

If a parallelepiped is cut from the crystal and the crystal planes are parallel to the $z-z$, $x-x$ and $y-y$ axes respectively, the crystal plate does not exhibit electrical properties in normal state. When an external force is applied on this parallelepiped, an electric field is formed along the $x-x$ direction, and its charge distribution is perpendicular to the plane of the $x-x$ axis, as shown in Figure 2-22. The longitudinal piezoelectric effect is generated by the force along the $x$ axis, the transverse piezoelectric effect is generated by the force along the $y$ axis, and the tangential piezoelectric effect is generated by the force along the relative two planes.

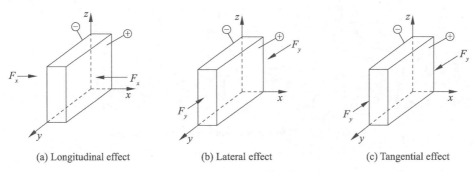

(a) Longitudinal effect　　　(b) Lateral effect　　　(c) Tangential effect

**Figure 2-22  Piezoelectric effect model**

## 2.3.1.2  Piezoelectric sensors

On the two working surfaces of the piezoelectric crystal plate, metal evaporation is carried out to form two electrodes ( metal film), as shown in Figure 2-23a. When the crystal plate is subjected to external force, an equal amount of charges with opposite polarities are accumulated on the two plates to form an electric field. Therefore, the piezoelectric sensor can be regarded as a charge generator, and can also constitute a capacitor. Its capacitance is still applicable to Equation (2-20).

If the external force applied to the crystal plate does not change, the charge accumulated on the plates has no internal leakage, and the external circuit load is infinite. Then the amount of charge remains constant during the external force, and the charge does not disappear until the external force is terminated. If the load is not infinite, the circuit will be discharged, and the charge on the plates cannot be kept constant, resulting in measurement error. Therefore, when using a piezoelectric sensor to measure static or quasi-static quantities, a very high impedance load must be used to reduce the leakage of charge. In dynamic measurement, the charge can be supplemented, and the amount of leakage charge is relatively small, so the piezoelectric sensor is suitable for dynamic measurement.

In actual piezoelectric sensors, two or more crystal plates are often used in parallel or series. Figure 2-23b shows the parallel connection. The negative electrodes of the two piezoelectric crystal plates are concentrated on the middle electrode plate, and the positive electrodes on both sides are connected in parallel. At this time, the capacitance is large, and the output charge is large. It is suitable for measuring the slowly varying signal and the occasion where the electric charge is the output. Figure 2-23c shows the series connection. The positive charge is concentrated on the upper plate and the negative charge is concentrated on the lower plate. The positive and negative electrodes of the two crystal plates in the middle are connected. At this time, the sensor capacitance is small, and the output voltage is large. It is suitable for the occasions where the voltage is the output signal.

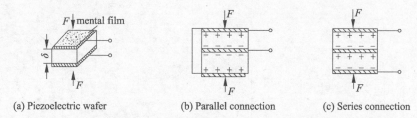

Figure 2-23　Piezoelectric wafer and its connection

(a) Piezoelectric wafer　　　(b) Parallel connection　　　(c) Series connection

### 2.3.1.3　Application of piezoelectric sensors

Figure 2-24 shows the working principle of a piezoelectric acceleration sensor. It is mainly composed of a base, piezoelectric crystal plate, mass block and spring. The base is fixed on the measured object. The vibration of the base causes the mass to generate an inertial force, which is opposite to the acceleration direction of the vibration. The inertial force acts on the piezoelectric wafer, making the surface of the piezoelectric wafer produce alternating voltage, which is proportional to the acceleration. After processing by the measuring circuit, it can be known the magnitude of acceleration.

Figure 2-25 shows the compressed vibration sensor for measuring abnormal vibration in automobile safety system. The sensor is mainly composed of piezoelectric crystal plate, mass block and spring. The mass block is pressed on the piezoelectric crystal plate by spring. When the car is running in the normal state, the mass vibrates keep the piezoelectric chip in a normal state of charge output. When the car is running under load, it will cause abnormal vibration or vibration caused by other noises, which leads to abnormal vibration of the mass block. The electrical signal of abnormal vibration of the car can be obtained from the test system.

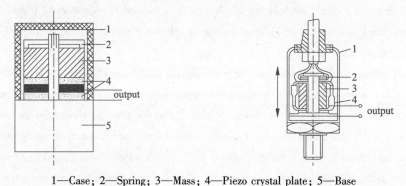

1—Case; 2—Spring; 3—Mass; 4—Piezo crystal plate; 5—Base

Figure 2-24　Piezoelectric accelerometer　　　Figure 2-25　Compression type vibration sensor

## 2.3.2　Magneto-electric sensor

The basic working principle of magneto-electric sensor is to convert the measured physical quantity into the induced electromotive force through magneto-electric effect. Magneto-electric sensor includes magneto-electric induction sensor, Hall sensor, etc.

## 2.3.2.1   Magneto-electric induction sensors

Magneto-electric induction sensor is also called induction sensor. It is a machine-electric energy conversion sensor. Magneto-electric sensor does not require external power supply, which has simple circuit, stable performance, small output impedance, and has a large frequency response range. It is suitable for testing of mechanical quantity, such as vibration, speed, torque, etc., but the size and weight of this kind of sensors are relatively large.

Based on Faraday Law of electromagnetic induction, when a coil with $W$ turns moves in a magnetic field to cut the magnetic line of induction, or the magnetic flux of the magnetic field where the coil is located changes, the magnitude of the induced electromotive force $e$ generated in the coil depends on the change rate of the magnetic flux $\Phi$ passing through the coil. That is

$$e = -W \frac{\mathrm{d}\Phi}{\mathrm{d}t} \qquad (2\text{-}30)$$

The change rate of magnetic flux is related to the strength of the magnetic field, the resistance of the magnetic circuit and the speed of the coil. Therefore, if one of these factors is changed, the induced electromotive force of the coil will be changed. According to different working principles, magneto-electric induction sensors can be divided into moving coil type and magneto-resistive type.

(1) Moving coil magneto-electric sensors

The moving coil magneto-electric sensor can be divided into linear velocity type and angular velocity type. Figure 2-26a shows the working principle of the linear velocity sensor. The movable coil is placed in a DC magnetic field generated by a permanent magnet. When the coil moves linearly in the magnetic field, the induced electromotive force generated is

$$e = WBlv\sin\theta \qquad (2\text{-}31)$$

In the equation, $B$ is the magnetic induction strength of the magnetic field; $l$ is the effective length of the single-turn coil; $W$ is the number of turns of the coil; $v$ is the moving speed of the coil relative to the magnetic field; $\theta$ is the angle between the direction of the coil movement and the direction of the magnetic field.

Figure 2-26b shows the working principle of the angular velocity sensor. The induced electromotive force generated when the coil rotates in the magnetic field is

$$e = kWBA\omega \qquad (2\text{-}32)$$

In the equation, $k$ is the structural coefficient of the sensor; $A$ is the average cross-sectional area of a single-turn coil; $\omega$ is the angular velocity.

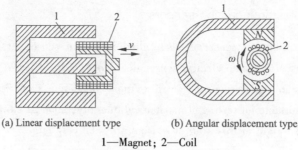

(a) Linear displacement type        (b) Angular displacement type

1—Magnet; 2—Coil

**Figure 2-26　Moving coil magneto-electric sensor**

In the magneto-electric sensor, when these structural parameters, such as $B$, $l$, $W$, $A$, are all constant values, the induced electromotive force is proportional to the speed of the coil's relative magnetic field ($v$ or $\omega$). So the basic form of this type of sensor is the speed sensor, which can directly measure linear or angular velocity. If the integration or differentiation circuit is connected to the measurement circuit, it can also be used to measure displacement or acceleration. Obviously, a magneto-electric induction sensor is only suitable for dynamic measurement.

(2) Magneto-resistive sensors

The coil and magnet of magneto-resistive sensors do not move relative to each other. The moving object (magnetically conductive material) changes the magnetic resistance of the circuit, causing the magnetic field lines to increase or weaken which make the coil generate the induced electromotive force. Its working principle and application example are shown in Figure 2-27. This kind of sensor consists of permanent magnet and coil wound around it.

For example, as shown in Figure 2-27a, the frequency of the gear can be measured. When the gear rotates, the convex and concave of the teeth cause the magnetic resistance to change, which causes the magnetic flux to change. The frequency of the induced AC electromotive force in the coil is equal to the product of the number of teeth and the speed of the measuring gear. The magneto-resistive magneto-electric sensor is easy to use and simple in structure. It can also be used to measure speed, eccentricity, vibration, etc. under different occasions.

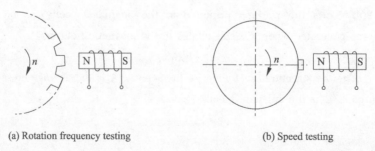

(a) Rotation frequency testing        (b) Speed testing

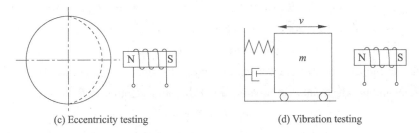

(c) Eccentricity testing                              (d) Vibration testing

**Figure 2-27    Working principle and application example of magneto-resistive sensor**

### 2.3.2.2    Hall sensors

The Hall sensor is a kind of magneto-electric sensor based on the Hall effect. The Hall sensor made of semiconductor has the characteristics of high sensitivity to magnetic field, simple structure and convenient to use. It is widely used to measure physical quantities, such as linear displacement, angular displacement and pressure.

(1) Hall effect and elements

The Hall element is a semiconductor magneto-electric conversion element. Generally, it is made of germanium (Ge), indium antimonide (InSb), indium arsenide (InAs) and other semiconductor materials. It works on the basis of the Hall effect. As shown in Figure 2-28a, place the Hall element in the magnetic field $B$. If the current $I$ is passed through the leads $a$ and $b$, then there will be a potential difference $e_H$ between the ends of $c$ and $d$. This phenomenon is called the Hall effect.

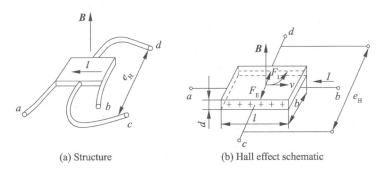

(a) Structure                              (b) Hall effect schematic

**Figure 2-28    Hall element and Hall effect principle**

The Hall effect is caused by the magnetic force acting on the moving charge. As shown in Figure 2-28b, assuming that the sheet is an N-type semiconductor, the direction of the magnetic field with magnetic induction intensity $B$ is perpendicular to the sheet, and the control current $I$ connects the left and right ends of the sheet, then the carriers (electrons) in the semiconductor will move in the opposite direction to the current $I$.

Due to the effect of the external magnetic field $B$, the electrons are deflected by magnetic field force $F_L$ (Lorentz force). As a result, electrons accumulate on the back face of the semiconductor, making it form negative charge, while the front face possesses positive charge due to lack of electrons, and an electric field is formed between the front and back faces. The electric field force $F_E$ generated by this electric field prevents the electrons from continuing to deflect.

When $F_E$ is equal to $F_L$, the electron accumulation reaches a dynamic equilibrium. At this time, the electric field between the front and back ends of the semiconductor ( that is perpendicular to the direction of the current and magnetic field) is called the Hall electric field, and the corresponding electromotive force is called the Hall electromotive force $e_H$. That is

$$e_H = K_H IB \sin \alpha \qquad (2\text{-}33)$$

In the equation, $K_H$ is the Hall constant, which is related to the physical properties and geometric dimensions of the current-carrying material, indicating the magnitude of hall electromotive force in unit magnetic induction strength and unit control current; $\alpha$ is the angle between the direction of the current and the magnetic field.

It can be seen that if you change $B$ or $I$ or both, the magnitude of the Hall electromotive force can be changed. With this characteristic, the measured mechanical quantity can be converted into the change of voltage.

(2) Application

Figure 2-29 shows the working principle of Hall effect displacement sensor. When the Hall element is placed in magnetic field, the direction of the magnetic field in the left half is upward, the direction of the magnetic field in the right half is downward and the current $I$ is input from the end of $a$. According to the Hall effect, Hall electromotive force $e_{H1}$ is generated in the left half, and the Hall electromotive force $e_{H2}$ in the opposite direction is generated in the right half. Therefore, the electromotive force between two ends of $c$ and $d$ is $e_{H1} - e_{H2}$. If the Hall element is in the initial position, there is $e_{H1} = e_{H2}$, then the output is zero. When the relative position of the magnetic polar system and the Hall element is changed, the displacement can be reflected by the change of the output voltage.

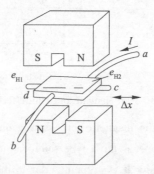

**Figure 2-29　Hall effect displacement sensor**

Figure 2-30 shows the working principle of detecting wire rope damage using Hall elements. In the figure, the permanent magnet makes the wire rope magnetized locally. When the wire rope is broken, a leakage magnetic field appears at the fracture. The Hall element will obtain a pulsating voltage signal when it passes through this magnetic field. After this signal is amplified, filtered, and A/D converted, it is analyzed by computer to identify the number of broken wires and the position of the port. At present, this technological achievement has been successfully used in the detection of broken wires in mine hoisting wire ropes and has obtained good benefits.

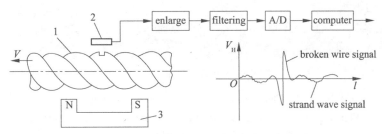

1—Wire rope; 2—Hall element; 3—Permanent magnet

**Figure 2-30   Detection principle of wire rope broken**

### 2.3.3   Photoelectric sensor

The photoelectric sensor is a sensor that converts light signals into electrical signals based on the photoelectric effect. The advantages of the photoelectric sensors are fast response speed, non-contact measurement, good accuracy, high resolution and good reliability. Therefore, it is an important sensitive device with a wide range of applications.

#### 2.3.3.1   Photoelectric effect and photosensitive element

Photoelectric effect refers to a physical phenomenon that some substance is irradiated with light and the electrical properties of the substance change. It can be divided into three types: external photoelectric effect, photoconductive effect and photovoltaic effect. According to these effects, different photoelectric converters can be made, which are collectively referred to as photosensitive elements.

(1) External photoelectric effect

The phenomenon in which light shines on certain substances (metals or semiconductors) to cause electrons to escape from the surface of these substances is called the external photoelectric effect. There are many types of photoelectric elements made based on the external photoelectric effect, the most typical one is the vacuum photoelectric cell, as shown in Figure 2-31. There are two electrodes in a vacuum glass bulb, a photocathode and a photoanode. The photocathodes are usually made of photosensitive materials (such as cesium CS). When light hits the photosensitive material, electrons escape. These electrons are attracted by the anode with positive potential, forming a space electron flow in the photoelectric tube, and generating current in the external circuit. If a resistor with a certain resistance is connected in series to the external circuit, the voltage on the resistor will vary with the intensity of the light, so as to achieve the purpose of converting the optical signal into an electrical signal.

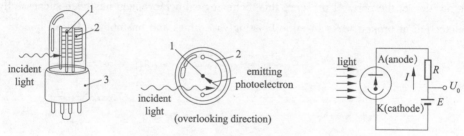

1—Photo anode; 2—Photo cathode; 3—Plug

**Figure 2-31　Structure and working principle of vacuum photocell**

(2) Photoconductive effect

After the semiconductor material is irradiated by light, the atoms inside it release electrons, but these electrons do not escape from the surface of the object and still remain inside, causing the resistivity of the object to change. This phenomenon is called the photoconductive effect. The photoresistor work based on this effect. As shown in Figure 2-32, the resistance value of the photosensitive resistor decreases when it is irradiated by

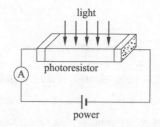

**Figure 2-32　Working principle of photoresistor**

light. In essence, due to the effect of light quantum, the photoresistor absorbs energy and releases electrons inside, which increases the carrier density or mobility, leading to the increase in conductivity and the decrease in resistance. The photoresistor is a kind of resistive device, which must be biased when used. When there is no light, its resistance is large and the current in the circuit is very small; when it is exposed to light, its resistance decreases and the circuit current increases rapidly.

(3) Photovoltaic effect

The photovoltaic effect refers to the phenomenon that the semiconductor material produces the electromotive force in a certain direction after the action of light shining. Semiconductor photovoltaic cell is the commonly used photovoltaic element, which directly converts light energy into electrical energy. When exposed to light, it directly outputs electric potential, which is actually equivalent to a power source. Figure

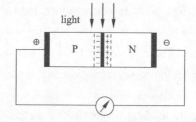

**Figure 2-33　Working principle of photovoltaic cell**

2-33 shows the principle of the photovoltaic cell with the PN junction. When irradiated by light, electrons and holes are generated near the PN junction due to the absorption of photon energy.

Under the electric field action of the PN junction, they produce drifting motion. The electrons are pushed to the N zone, and the holes are pulled into the P zone. As a result, a large number of excess holes are accumulated in the P zone, and a large number of excess electrons are accumulated in the N zone, making the P zone carry positive electricity and the N zone carry negative electricity. There is a potential difference between the P zone and N zone, and there is current flowing in the circuit after connecting with the wire. Commonly used photovoltaic cell materials are selenium, silicon, cadmium telluride, cadmium sulfide, etc. Among them, silicon photovoltaic cells are the most widely used, which are simple and portable, do not produce gas or thermal pollution, easy to adapt to the environment, especially suitable for providing power for various instruments of spacecraft.

### 2.3.3.2   Application of photoelectric sensors

Photoelectric sensors are widely used in the field of mechanical engineering. Here are some examples to illustrate the specific applications of photoelectric sensors.

Figure 2-34 is the use of photoelectric sensor to detect the surface defects of the work-piece. The light beam emitted by the laser tube 1 becomes a parallel light beam through lenses 2 and 3, and the parallel light beam is focused on the surface of the work-piece 7 by the lens 4 to form a slender light band with a width of about 0.1 mm. The diaphragm 5 is used to control the luminous flux. If there are defects (non-circular, rough, crack, etc.) on the surface of the work-piece, it will cause the deflection on scattering of the light beam, which can be received by the photocell 6 and can be output by converting into electrical signal.

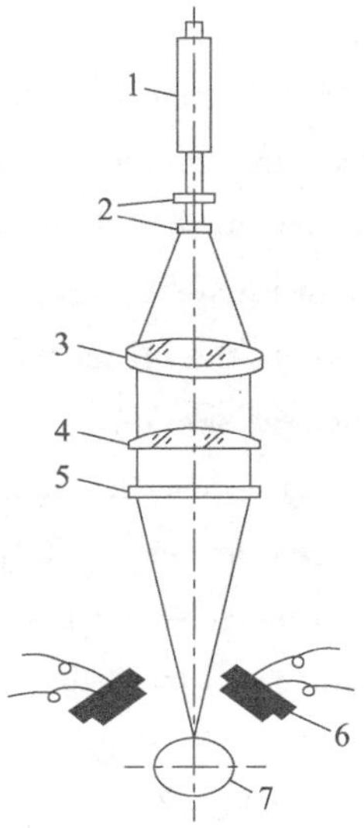

1—Laser tube; 2,3,4—Lens; 5—Iris; 6—Photoelectric sensor; 7—Work-piece

**Figure 2-34   Detect surface detects of workpiece by photoelectric sensor**

Figure 2-35 shows the working principle of the photoelectric tachometer. The rotating shaft of the motor is painted with black and white colors. When the motor is rotating, the reflected light and the non-reflected light appear alternately. The photoelectric element receives the light reflection signal intermittently, outputs the electrical signal intermittently, and then outputs the square wave signal through the amplifying and shaping circuit. Finally, the speed of the motor is measured by the digital frequency meter.

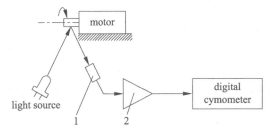

1—Photoelectric element; 2—Amplification and shaping circuit

**Figure 2-35   Working principle of photoelectric tachometer**

In recent years, sensor technology has developed rapidly, and new varieties, new structures and new applications of sensors have emerged continuously, such as wireless sensors, intelligent sensors, biosensors, etc. Interested readers can further read the relevant information of sensors.

## 2.4　Select the appropriate sensor

In the actual mechanical quantity test, how to choose the sensor reasonably according to the test purpose and actual conditions is a difficult problem that we need to solve first. Therefore, on the basis of the preliminary knowledge of sensor principle mentioned above, this section introduces some main technical indicators that should be considered when selecting sensors.

(1) Sensitivity

In general, the higher the sensitivity of the sensor, the better, so that even if there is only a small change in the test, the sensor can have a large output. However, it should also be considered that when the sensitivity is high, interference signals that are not related to the measured signal are more likely to be mixed in and will be amplified by the amplification system. Therefore, when selecting a sensor, not only need to consider ensuring high sensitivity, but also the noise generated by itself is small, and it is not easy to be interfered by the outside during the test process, that is, the sensor is also required to have a higher signal-to-noise ratio.

The sensitivity of a sensor is closely related to its measuring range. When measuring, unless there is an accurate nonlinear correction method, it is necessary to ensure that the input does not exceed the measurement range of the sensor. In the actual measurement, the input includes not only the measured signal, but also the interference signal. Therefore, if the sensitivity is selected too high, the measurement range of the sensor will be affected.

(2) Response characteristics

The actual sensor always has a certain time delay, and we generally hope that the time delay is as small as possible.

Generally speaking, physical sensors based on photoelectric effect and piezoelectric effect of materials have fast response speed and wide working frequency range. However, structural sensors, such as inductors, capacitors, magnetoelectric sensors, etc. are limited by the inertia of mechanical system, so their natural frequency and working frequency are low.

In dynamic measurement, the response characteristics of the sensor have a direct impact on the measurement results. So the sensor should be selected reasonably according to its response characteristics and the concrete measured signal (such as steady-state, transient or random signal).

(3) Linear measurement range

The sensor has a certain linear range, in which the output is proportional to the input. The wider the linear range, the larger the measurement range of the sensor.

The basic condition of accurate measurement is that the sensor works in the linear range. For example, the elastic limit of the force measuring element in a mechanical sensor is a fundamental factor that determines the force range. When the elastic limit is exceeded, a linear error will occur.

However, it is difficult for any sensor to guarantee its absolute linearity. Within the allowable error range, it can be applied in its approximate linear region. For example, variable-gap capacitive and inductive sensors work in the approximate linear region near the initial gap. Therefore, in engineering application, when selecting a sensor, the variation range of the measured signal must be considered to make its non-linear error within the allowable range.

(4) Stability

After a long period of use, the sensor should also have the performance to keep its original output characteristics unchanged, that is, high stability. In order to ensure high stability in sensor application, sensors with good design and manufacture and suitable use conditions shall be selected in advance. Meanwhile, the prescribed operating conditions should be strictly maintained in the process of use to reduce the adverse effects of the operating conditions as far as possible. For example, the dust on the surface of a potentiometer sensor will bring additional noise to the input of the sensor. For another example, the capacitance sensor with variable gap, the ambient humidity or the oil immersed in the gap, will change the dielectric constant of the medium. When magnetoelectric sensor or Hall element works in electric field or magnetic field, it will bring measurement errors. When there is dust or water vapor on the sensitive surface of photoelectric sensor, the luminous flux and spectral composition will be changed.

In mechanical engineering, some mechanical systems or automatic machining processes require that sensors can be used for a long time without frequent replacement or calibration. For example, the force measuring system of the adaptive grinding process or the automatic measuring device of part size, etc. In this case, the stability of the sensor should be fully considered.

(5) Accuracy

The accuracy of the sensor reflects the consistency between the output of the sensor and the measured signal. The sensor is located at the input of the test system. Therefore, whether the sensor can truly reflect the measured signal has a direct impact on the entire test system.

However, not only the accuracy of the sensor is required to be as high as possible, and economy should also be considered. The higher the accuracy of the sensor, the more expensive it is. Therefore, the selection should be achieved based on the measurement purpose and the cost performance of the test system, and the specific conditions should be analyzed in detail. Aiming at the qualitative measurement or comparative research, it is not required to measure absolute value, the accuracy requirements of the sensor can be appropriately reduced. When the signal is to be quantitatively analyzed, the sensor is required to have a sufficiently high accuracy.

(6) Measurement method

Another important factor that needs to be considered when selecting a sensor is how it works in practical applications, such as contact measurement and non-contact measurement, online measurement and off-line measurement, etc. Different working modes of sensors have different requirements for sensors, so it should be fully considered in selection.

(7) Other aspects

In addition to the above factors that should be fully considered, the conditions of simple structure, small size, light weight, cost-effective, easy maintenance and replacement should also be considered in sensors selection.

# Questions

2.1 Briefly describe the definition and classification of sensors.

2.2 What is the resistance strain effect of metals? What is the physical significance of wire sensitivity? What are the characteristics?

2.3 What is the piezoresistive effect of semiconductors? What are the characteristics of the sensitivity of semiconductor strain gauges?

2.4 Explain the basic working principle and advantages of eddy current sensors.

2.5 Briefly describe the working principle and classification of capacitive sensors.

2.6 The radius of the circular plate of a capacitive sensor (parallel plate capacitor) is $R = 4$ mm, the initial distance between the plates is $\delta_0 = 0.3$ mm, and the medium is air. Try to find:

(1) If the change in the distance between the plates is $\Delta\delta = \pm 1$ μm, what is the change in capacitance $\Delta C$?

(2) If the sensitivity of the measuring circuit is $K_1 = 100$ mV/pF and the sensitivity of the reading instrument is $K_2 = 5$ grid/mV, what is the change of the reading instrument at $\Delta\delta = \pm 1$ μm?

2.7 What is the piezoelectric effect? Why are piezoelectric sensors commonly used to measure dynamic signals?

2.8 Explain the basic working principle and structure of the magnetoelectric sensor.

2.9 What is the Hall effect? What is its physical nature? What physical quantities can be measured with Hall components? Please give three examples for illustration.

2.10 What is the photoelectric effect? What are the types? What are the corresponding optoelectronic components?

2.11 Briefly describe the working principle of ultrasonic straight probe and CCD image sensor, and give examples to illustrate their respective applications.

2.12 Try to distinguish between contact and non-contact sensors, list their names, transformation principles, and where are they used?

2.13 There are a number of turbine blades that need to be tested for cracks. Please list two or more methods and explain the principle of the sensors used.

2.14 What are the basic principles for selecting sensors? How should these principles be considered in the application? Give an example to illustrate.

# Chapter 3

## Basic Characteristics of Test System

The general test system consists of three parts: sensor, intermediate conversion device and display recording device. During the test, the physical quantities (such as pressure, acceleration, temperature, etc.) that reflect the characteristics of the measured object are detected by sensors and converted into electrical signals, and then transmitted to the intermediate conversion device. In the intermediate conversion device, the electrical signal is processed by the hardware circuit or converted into a digital quantity by A/D circuit, and then the result obtained is transmitted to the display recording device in the form of an electrical signal or a digital signal. Finally, the measurement results are displayed by the display and recording device and provided to the observer or other automatic control devices. The test system is shown in Figure 3-1.

According to the complexity of the test task, each part in the test system can be composed of multiple modules. For example, the intermediate transformation device in the machine tool bearing fault monitoring system shown in Figure 3-2 is composed of three modules: a band-pass filter, an A/D converter, and Fast Fourier Transform (FFT) analysis software. The sensor in the test system is a vibration accelerometer, which converts the vibration signal of the machine tool bearing into an electrical signal; the band-pass filter is used to filter out high and low frequency interference signals and to amplify the signal, A/D converter is used to sample the amplified test signal and convert it into digital quantity; the FFT analysis software performs fast Fourier transform on the converted digital signal and calculates the frequency spectrum of the signal; finally, the spectrum is displayed by a computer display.

To achieve testing, a test system must be reliable and undistorted. Therefore, this chapter will discuss the relationship between the test system input and its output, as well as the conditions of the test system without distortion.

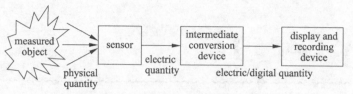

**Figure 3-1　Sketch of the test system**

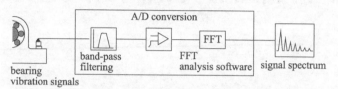

**Figure 3-2　Test system for bearing vibration signal**

# 3.1　Linear system and its basic properties

　　The essence of mechanical testing is to study the relationship among the signal (excitation) of the machine under test, the characteristics of the test system and the test results (response), which can be shown in Figure 3-3.

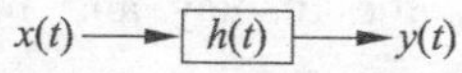

$$x(t) \longrightarrow \boxed{h(t)} \longrightarrow y(t)$$

**Figure 3-3　Relationship among the test system, the input and output**

　　There are three meanings:

　　① If the input $x(t)$ and output $y(t)$ are measurable, the characteristic $h(t)$ of the test system can be inferred;

　　② If the characteristic $h(t)$ of the test system is known and the output $y(t)$ is measurable, the corresponding input $x(t)$ can be derived;

　　③ If the input $x(t)$ and system characteristics $h(t)$ are known, the output $y(t)$ of the system can be inferred or estimated.

　　The test system mentioned here, in a broad sense, refers to the entire system from a certain excitation input to the detection output, and generally includes two parts: the device under test and the measuring device. Therefore, only by first ascertaining the characteristic of the measuring device can the characteristic or running state of the tested device be evaluated correctly from the test results.

　　The ideal test device should have a single-valued, determined input/output relationship, and preferably a linear relationship. Since the correction and compensation techniques are easy to implement in static measurement, this linear relationship is not necessary (but desirable); however, in dynamic measurement, the test device should strive to be a linear system. There are two main reasons: first, the current mathematical processing and analysis methods for linear systems are relatively perfect; second, it is difficult to correct nonlinearities in dynamic measurement. However, for many actual mechanical signal testing devices, it is impossible to maintain linearity in a large working range, and they can only be treated as a linear system within a certain working range and tolerance range.

The relationship between input $x(t)$ and output $y(t)$ of a linear system can be described by Equation (3-1).

$$a_n \frac{\mathrm{d}^n y(t)}{\mathrm{d}t^n} + a_{n-1} \frac{\mathrm{d}^{n-1} y(t)}{\mathrm{d}t^{n-1}} + \cdots + a_1 \frac{\mathrm{d}y(t)}{\mathrm{d}t} + a_0 y(t)$$
$$= b_m \frac{\mathrm{d}^m x(t)}{\mathrm{d}t^m} + b_{m-1} \frac{\mathrm{d}^{m-1} x(t)}{\mathrm{d}t^{m-1}} + \cdots + b_1 \frac{\mathrm{d}x(t)}{\mathrm{d}t} + b_0 x(t) \tag{3-1}$$

When $a_n$, $a_{n-1}$, $\cdots$, $a_0$ and $b_m$, $b_{m-1}, \cdots, b_0$ are all constants, Equation (3-1) describes a linear system, also called a time-invariant linear system, it has the following main basic properties:

① Superposition

If $x_1(t) \rightarrow y_1(t)$, $x_2(t) \rightarrow y_2(t)$, then there is

$$[x_1(t) \pm x_2(t)] \rightarrow [y_1(t) \pm y_2(t)] \tag{3-2}$$

② Proportionality

If $x(t) \rightarrow y(t)$, then there is the following equation for any constant

$$cx(t) \rightarrow cy(t) \tag{3-3}$$

③ Differentiation

If $x(t) \rightarrow y(t)$, then there is

$$\frac{\mathrm{d}x(t)}{\mathrm{d}t} \rightarrow \frac{\mathrm{d}y(t)}{\mathrm{d}t} \tag{3-4}$$

④ Integration

If the initial state of the system is zero, $x(t) \rightarrow y(t)$, then there is

$$\int_0^t x(t)\,\mathrm{d}t \rightarrow \int_0^t y(t)\,\mathrm{d}t \tag{3-5}$$

⑤ Frequency retention

If the input of the system is sinusoidal signal at a certain frequency, the steady-state output of the system will only be at the same frequency.

Suppose the input of the system is a sinusoidal signal: $x(t) = x_0 e^{\mathrm{j}\omega_0 t}$, then the steady-state output of the system is

$$y(t) = y_0 e^{\mathrm{j}(\omega_0 t + \varphi)} \tag{3-6}$$

The characteristics of the linear system mentioned above, especially the frequency retention, play an important role in the test work. Because in the actual test, the measured signal is often interfered by other signals or noise. At this time, based on the frequency retention characteristic, it can be concluded that only the same frequency components as the input signal in the measured signal are the output caused by the input. Similarly, in mechanical fault diagnosis, according to the main frequency component of the test signal, on the basis of eliminating interference, the input signal should also contain this frequency component according to the frequency retention characteristics. By looking for the cause of the frequency component, the cause of the failure can be diagnosed.

# 3.2 Static characteristics of test system

The static characteristic of the test system is to describe the closeness of the actual test system to the ideal linear time-invariant system under the condition of static quantity measurement. When measuring static quantities, the response characteristics shown by the device are called static response characteristics. The parameters commonly used to describe static response characteristics mainly include sensitivity, nonlinearity, and return error.

## 3.2.1 Sensitivity

When the input $x(t)$ of the test system has a $\Delta x$ at a certain time $t$, and the output $y$ has a corresponding change $\Delta y$ when it reaches the new steady state, then

$$S = \frac{\Delta y}{\Delta x} \tag{3-7}$$

It is the absolute sensitivity of the test system, as shown in Figure 3-4a.

If the transition process of the system is ignored, it can be seen from the nature of the linear system that the sensitivity of the linear system can be expressed as

$$S = \frac{b_0}{a_0} = C \tag{3-8}$$

In the equation, $a_0$ and $b_0$ are constants, and $C$ represents a proportional constant.

It can be seen that the static characteristic curve of the linear system is a straight line. For example, when the displacement change is 1 μm, the output voltage of the displacement test system is 5 mV, then its sensitivity is $S = 5$ V/mm. For test systems with the same input and output dimensions, the sensitivity is dimensionless, often called the magnification.

Due to the change of external environmental conditions and other factors, the characteristics of the output in test system may change. For example, changes in amplifier circuit characteristics caused by changes in ambient temperature are ultimately reflected as changes in sensitivity. The resulting changes in sensitivity are called sensitivity drift, as shown in Figure 3-4b.

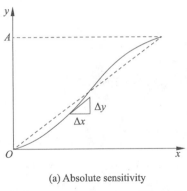

(a) Absolute sensitivity

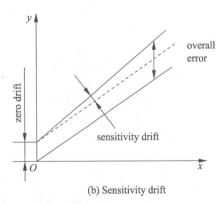

(b) Sensitivity drift

**Figure 3-4   Absolute sensitivity and its drift**

When designing or selecting the sensitivity of the test system, it should be carried out reasonably according to the test requirements. Generally speaking, the higher the sensitivity of the test system, the narrower the test range and the poorer the stability.

## 3.2.2   Nonlinearity

Nonlinearity is a static index that measures the linearity between the input and output of the test system. Usually, the input-output relation curve of the test system (namely, the experimental curve or the calibration curve) is obtained by static test experiment. The deviation of the curve from its fitted straight line is the nonlinearity. The nonlinearity $F$ can be defined as the ratio of the maximum deviation $B$ of the experimental curve to the fitted straight line to the output range $A$ within the full test range. As shown in Figure 3-5.

$$F = \frac{B_{max}}{A} \times 100\% \tag{3-9}$$

## 3.2.3   Return error

The reason for the return error is generally due to the lag or dead zone in the test system. It is also an index to characterize the nonlinear characteristics of the test system, which can reflect the situation that the same input corresponds to several different outputs, and is usually obtained by static measurement, as shown in Figure 3-6. It is defined as: under the same test conditions, within the whole test range, when the input quantity increases slightly or decreases greatly, the ratio of the maximum value (the difference between two output values for the same input quantity) to the entire output range.

$$H = \frac{h_{max}}{A} \times 100\% \tag{3-10}$$

The return error may be caused by friction, clearance, deformation under force of the material, hysteresis and other factors, or it may reflect the non-working area (also known as dead area) of the instrument. The so-called non-working area is the range where input changes have no effect on the output.

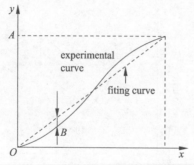 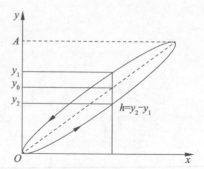

**Figure 3-5  Nonlinearity of test system**          **Figure 3-6   Return error of test system**

# 3.3   Dynamic characteristics of test system

The dynamic characteristics of the test system refer to the relationship between the output changes and the input when the input changes with time. Generally, within the measurement range under consideration, the test system can be regarded as a linear system. Therefore, Equation (3-1), a time-invariant linear system differential equation, can be used to describe the relationship among the test system, the input and output, but there are many inconveniences when using it. Therefore, the "transfer function" of its response is often established by the Laplace transform, and the corresponding "frequency response function" is established by the Fourier transform, so as to describe the characteristics of the test system more easily. The content of this section is difficult. For more details, please refer to the relevant knowledge of "Mechanical Engineering Control Fundamentals".

## 3.3.1   Theoretical basis of dynamic characteristics

### 3.3.1.1   Transfer function

When measuring the operating machinery, the measurement results obtained are not only affected by the static characteristics of the equipment, but also by the dynamic characteristics of the test system. Therefore, a clear understanding of the dynamic characteristics of the test system is required. The relationship between input and output in the test system is described in Equation (3-1). For a linear system, if the initial condition of the system is zero, that is, before the observation time $t(t \rightarrow 0^-)$, its input and output signals and their derivatives are all zero, then the Laplace transform of Equation (3-1) can be obtained

$$(a_n s^n + a_{n-1} s^{n-1} + \cdots + a_1 s + a_0) Y(s) = (b_m s^m + b_{m-1} s^{m-1} + \cdots + b_1 s + b_0) X(s) \qquad (3-11)$$

Define the ratio of the Laplace transform of the output signal and the input signal as the transfer function, namely

$$H(s) = \frac{Y(s)}{X(s)} = \frac{b_m s^m + b_{m-1} s^{m-1} + \cdots + b_1 s + b_0}{a_n s^n + a_{n-1} s^{n-1} + \cdots + a_1 s + a_0} \qquad (3\text{-}12)$$

In the equation, $s$ is the Laplace operator, $s = \alpha + j\omega$; $a_n, a_{n-1}, \cdots, a_1, a_0$ and, $b_n, b_{n-1}, \cdots, b_1, b_0$ are constant coefficients determined by the physical parameters of the test system.

It can be seen from Equation (3-12) that the transfer function represents the transmission and conversion characteristics of the system to the input signal in the form of an algebraic expression. It contains all the information about transient $s = \alpha$ and steady state $s = j\omega$ responses. Equation (3-1) represents the relationship between system input and output signals in the form of differential equations. In operation, the transfer function is easier than solving the differential equation. The transfer function has the following main characteristics:

① $H(s)$ describes the inherent dynamic characteristics of the system itself, and has nothing to do with input $x(t)$ and the initial state of the system.

② $H(s)$ is a mathematical description of the characteristics of a physical system, and has nothing to do with the specific physical structure of the system. After abstracting the actual physical system into a mathematical model Equation (3-1), $H(s)$ is obtained after Laplace transformation, so that the transfer function of the same form can represent different physical systems with the same transmission characteristics.

③ The denominator of $H(s)$ depends on the structure of the system, while the numerator represents the relationship between the system and the outside world, such as the position of the input point, input means and measured point layout, etc. The power of $s$ in the denominator represents the order of the differential equation of the system, such as when $n = 1$ or $n = 2$, which is respectively called first-order or second-order system.

A general test system is a stable system where the power of $s$ in the denominator is always higher than the power of $s$ in the numerator ($n > m$).

### 3.3.1.2 Frequency response function

The transfer function $H(s)$ is to describe and investigate the characteristics of the system in the complex number domain, and has many advantages compared with the time domain using differential equations to describe and investigate the characteristics of the system. The frequency response function is to describe and investigate the system characteristics in the frequency domain. Compared with the transfer function, the frequency response function is easy to be established through experiments, and its physical concept is clear.

In the case that the system transfer function $H(s)$ is already known, let the real part of $s$ in $H(s)$ be zero, that is, $s = j\omega$ can obtain the frequency response function $H(s)$. For time-invariant linear systems, there is a frequency response function $H(\omega)$

$$H(\omega) = \frac{b_m (j\omega)^m + b_{m-1} (j\omega)^{m-1} + \cdots + b_1 (j\omega) + b_0}{a_n (j\omega)^n + a_{n-1} (j\omega)^{n-1} + \cdots + a_1 (j\omega) + a_0} \qquad (3\text{-}13)$$

Where $j = \sqrt{-1}$.

If the input signal is connected to a time-invariant linear system at time $t=0$, and $s=\mathrm{j}\omega$ is substituted into the Laplace transform, the Laplace transform is actually transformed into a Fourier transform. And since the initial condition of the system is zero, the frequency response function $H(\omega)$ of the system becomes the ratio of $Y(\omega)$ and $X(\omega)$, namely

$$H(\omega)=\frac{Y(\omega)}{X(\omega)} \tag{3-14}$$

Where $Y(\omega)$ and $X(\omega)$ are the Fourier transform of output $y(t)$ and input $x(t)$ respectively. According to Equation (3-14), after measuring output $y(t)$ and input $x(t)$, the frequency response function $H(w)=Y(w)/X(w)$ can be obtained from their Fourier transform $Y(\omega)$ and $X(\omega)$. The frequency response function describes the relationship between the harmonic input of a system and its steady-state output, when measuring the system frequency response function, it must be measured when the system response reaches the steady state stage.

The frequency response function is complex, so it can be written in complex exponential form

$$H(\omega)=A(\omega)\,\mathrm{e}^{\mathrm{j}\varphi(\omega)} \tag{3-15}$$

Where $A(\omega)$ is called the amplitude-frequency characteristic of the system; $\varphi(\omega)$ is called the phase-frequency characteristic of the system.

It can be seen that the frequency response function $H(\omega)$ of the system or its amplitude-frequency characteristic $A(\omega)$ and phase-frequency characteristic $\varphi(\omega)$ are all functions of the harmonic input frequency $\omega$.

For the convenience of research, curves are sometimes used to describe the transmission characteristics of the system. Curve $A(\omega)-\omega$ is called the amplitude-frequency characteristic curve of the system. Curve $\varphi(\omega)-\omega$ is called the phase frequency characteristic curve. In actual drawing, the logarithmic scale is often used for the independent variables, and the amplitude coordinates are taken in decibels, that is, $20\lg A(\omega)-\lg \omega$ and $\varphi(\omega)-\lg \omega$ curves, which are respectively called logarithmic amplitude-frequency curve and logarithmic phase-frequency curve, collectively called Bode diagram.

If $H(\omega)$ is written in the form of real and imaginary parts, there are

$$H(\omega)=P(\omega)+\mathrm{j}Q(\omega)$$

In which, both $P(\omega)$ and $Q(\omega)$ are real functions of $\omega$, curves $P(\omega)-\omega$ and $Q(\omega)-\omega$ are respectively called the real frequency characteristic and imaginary frequency characteristic curve of the system. If the real part and imaginary part of $H(\omega)$ are taken as the ordinate and abscissa respectively, then curve $Q(\omega)-P(\omega)$ is called the Nyquist chart. Obviously

$$A(\omega)=\sqrt{P^2(\omega)+Q^2(\omega)} \tag{3-16}$$

$$\varphi(\omega)=\arctan\frac{Q(\omega)}{P(\omega)} \tag{3-17}$$

### 3.3.1.3 Impulse response function

If the input of the test system is a unit pulse function, that is, at $x(t)=\delta(t)$, there is $X(s)=1$. Therefore, there is

$$H(s) = \frac{Y(s)}{X(s)} = Y(s)$$

Take the inverse Laplace transform of the above equation, we have

$$y(t) = h(t)$$

Call $h(t)$ the impulse response function of the system. The impulse response function is a time domain description of the characteristics of the test system.

So far, the dynamic characteristics of the test system can be described by $h(t)$ in the time domain, $H(\omega)$ in the frequency domain, and $H(s)$ in the complex domain. The relationship of the three is one-to-one correspondence.

### 3.3.1.4　Series and parallel connection of parts

The actual test system usually consists of several parts, the relationship between the transfer function of the test system and the transfer function of each part depends on the connection form of each part. If the system is composed of several parts in series, as shown in Figure 3-7, and the latter part has no influence on the former part, and the transfer function of each part itself is $H_i(s)$, then the total transfer function of the test system is

$$H(s) = \prod_{i=1}^{n} H_i(s) \tag{3-18}$$

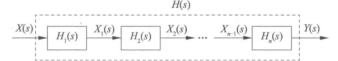

**Figure 3-7　System series connection**

The frequency response function of the corresponding system is

$$H(j\omega) = \prod_{i=1}^{n} H_i(j\omega) \tag{3-19}$$

Its amplitude-frequency and phase-frequency characteristics are

$$A(\omega) = \prod_{i=1}^{n} A_i(\omega)$$

$$\varphi(\omega) = \sum_{i=1}^{n} \varphi_i(\omega) \tag{3-20}$$

If the system is composed of multiple parts in parallel, as shown in Figure 3-8, the total transfer function of the test system is

$$H(s) = \sum_{i=1}^{n} H_i(s) \tag{3-21}$$

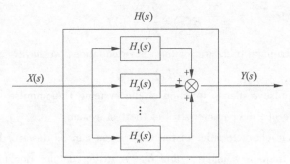

**Figure 3-8  System parallel connection**

The frequency response function of the corresponding system is

$$H(j\omega) = \sum_{i=1}^{n} H_i(j\omega) \tag{3-22}$$

Note: When the power value $n$ of $s$ in the denominator of the transfer function of the system is greater than 2, the system is called a higher-order system. Since the general test system is always stable, the denominator of the transfer function of the higher-order system can always be decomposed into the first and second real coefficient factors of $s$, namely

$$a_n s^n + a_{n-1} s^{n-1} + \cdots + a_1 s + a_0 = a_n \prod_{i=1}^{r} (s + p_i) \prod_{i=1}^{(n-r)/2} (s^2 + 2\zeta_i \omega_{ni} s + \omega_{ni}^2) \tag{3-23}$$

In the equation, $p_i$, $\zeta_i$, $\omega_{ni}$ are real constants, and $\zeta_i < 1$.

Therefore, Equation (3-12) can be rewritten as

$$H(S) = \sum_{i=1}^{r} \frac{q_i}{s + p_i} + \sum_{i=1}^{(n-r)/2} \frac{\alpha_i s + \beta_i}{s^2 + 2\zeta_i \omega_{ni} s + \omega_{ni}^2} \tag{3-24}$$

Where $\alpha_i$, $\beta_i$, $q_i$ are real constants.

Equation (3-24) shows that any high-order system can always be regarded as a series and parallel connection of several first-order and second-order systems. Therefore, studying the dynamic characteristics of first-order and second-order systems is of very general significance.

## 3.3.2  Dynamic analysis of first-order and second-order systems

It can be seen from the foregoing section 3.3.1 that the first-order system and the second-order system are the basis for the analysis and research of higher-order systems. Therefore, this section will introduce in detail the characteristics of the first and second order systems and their response under typical signal input.

### 3.3.2.1  Dynamic characteristics of first-order systems

Let us start with a concrete example. Figure 3-9 is a liquid column thermometer. $T_i(t)$ represents the input signal of the thermometer taken as the measured temperature, and $T_0(t)$ representing the output signal of the thermometer taken as the indicated temperature.

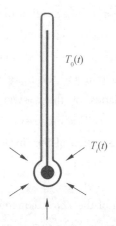

$T_0(t)$

$T_i(t)$

**Fig. 3-9　Liquid column thermometer**

The relationship between input and output is

$$\frac{T_i(t)-T_0(t)}{R}=C\frac{\mathrm{d}T_0(t)}{\mathrm{d}t} \tag{3-25}$$

$$RC\frac{\mathrm{d}T_0(t)}{\mathrm{d}t}+T_0(t)=T_i(t) \tag{3-26}$$

Where $R$ is the thermal resistance of the transmission medium; $C$ is the heat capacity of the thermometer.

　　Equation (3-26) shows that the differential equation of the liquid column thermometer system is a first-order differential equation, and the thermometer can be considered as a first-order test system. Laplace transform is applied to it, and set $\tau = RC$ ($\tau$ is the thermometer time constant), then

$$\tau s T_0(s)+T_0(s)=T_i(s) \tag{3-27}$$

　　Therefore, the transfer function is

$$H(s)=\frac{T_0(s)}{T_i(s)}=\frac{1}{1+\tau s} \tag{3-28}$$

　　Correspondingly, the frequency response function of the thermometer system is

$$H(\mathrm{j}\omega)=\frac{1}{1+\mathrm{j}\omega\tau} \tag{3-29}$$

It can be seen that the transfer characteristics of liquid column thermometers have first-order system characteristics.

　　The following analyzes the frequency response characteristics of the first-order system in a general sense. The general formula of the first-order system differential equation is

$$a_1\frac{\mathrm{d}y(t)}{\mathrm{d}t}+a_0y(t)=b_0x(t) \tag{3-30}$$

Divide the terms of the equation by $a_0$ to get

$$\frac{a_1}{a_0}\frac{\mathrm{d}y(t)}{\mathrm{d}t}+y(t)=\frac{b_0}{a_0}x(t) \tag{3-31}$$

In the formula, $a_1/a_0$ has the dimension of time, called time constant, which is often represented by the symbol $\tau$; $b_0/a_0$ is the static sensitivity of the system, denoted by $S$.

In a linear system, $S$ is a constant. Since the value of $S$ only represents the enlarged proportional relationship between the output and the input (when the input is a static quantity), the study of the dynamic characteristics of the system is not affected. Therefore, for the convenience of research, set $S = b_0/a_0 = 1$. This processing is called sensitivity normalization. After the above processing, the differential equation of the first-order system can be rewritten as

$$\tau \frac{dy(t)}{dt} + y(t) = x(t) \tag{3-32}$$

Let us take the Laplace transform of the above equation

$$\tau sY(s) + Y(s) = X(s) \tag{3-33}$$

Then the transfer function of the first-order system is

$$H(s) = \frac{Y(s)}{X(s)} = \frac{1}{\tau s + 1} \tag{3-34}$$

Its frequency response is

$$\begin{cases} H(j\omega) = \dfrac{1}{j\omega\tau} = \dfrac{1}{1+(\omega\tau)^2} - j\dfrac{\omega\tau}{1+(\omega\tau)^2} \\[2mm] A(\omega) = \sqrt{[\operatorname{Re}(\omega)]^2 + [\operatorname{Im}(\omega)]^2} = \dfrac{1}{\sqrt{1+(\omega\tau)^2}} \\[2mm] \varphi(\omega) = \arctan\dfrac{\operatorname{Im}(\omega)}{\operatorname{Re}(\omega)} = -\arctan(\omega\tau) \end{cases} \tag{3-35}$$

A negative value of $\varphi(\omega)$ indicates that the phase of the output signal of the system lags behind that of the input signal. The amplitude-frequency and phase-frequency characteristic curves of the first-order system are shown in Figure 3-10.

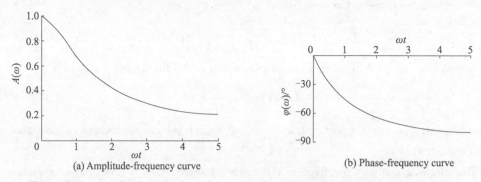

(a) Amplitude-frequency curve          (b) Phase-frequency curve

**Figure 3-10    Amplitude and phase frequency characteristics of the first-order system**

From the amplitude-frequency curve of the first-order system, compared with the condition of no distortion in the dynamic test, it is obvious that it does not meet the requirement that $A(\omega)$ is a horizontal straight line. For the actual test system, it is almost impossible to fully satisfy the theoretical non-distortion condition of dynamic test. It can only require that the amplitude error does not exceed a certain limit in a certain frequency range close to the non-distorted test condition. Generally, in the absence of specific precision requirements, the system is considered to be able to meet the dynamic test requirements as long as it works in the frequency range with the amplitude error less than 5% (that is, after the system sensitivity is normalized, the value of $A(\omega)$ is not greater than 1.05 or not less than 0.95). When the first-order system is $\omega = 1/\tau$, the value of $A(\omega)$ is 0.707 (−3 dB), the phase lags 45°, and $\omega = 1/\tau$ is usually called the corner frequency of the first-order system. Only when $\omega$ is much less than $1/\tau$, the amplitude-frequency characteristic is close to 1, and the dynamic test requirements can be satisfied to varying condition. In the case of a certain amplitude error, the smaller $\tau$ is, the larger the operating frequency range of the system is. In other words, when the highest frequency component $\omega$ of the measured signal is constant, the smaller $\tau$ is, the smaller the amplitude error of the system output is.

From the phase frequency curve of the first-order system, it can be seen that only when $\omega$ is much less than $1/\tau$, the phase frequency curve is close to a zero-crossing oblique line, which can meet the non-distortion condition of dynamic test to varying degrees. Also, the smaller $\tau$ is, the larger the operating frequency range of the system is.

Based on the above analysis, it can be concluded: the index parameter $\tau$ reflecting the dynamic performance of the first-order system is the time constant. In principle, the smaller $\tau$ is, the better.

In common measuring devices, spring damping system and simple RC low-pass filter are first-order systems, as shown in Figure 3-11.

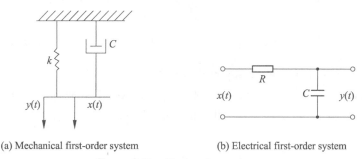

(a) Mechanical first-order system          (b) Electrical first-order system

**Figure 3-11    First order systems**

### 3.3.2.2   Dynamic characteristics of the second-order system

The moving coil type display vibrator shown in Figure 3-12 is a typical second-order system. In moving coil type vibrators such as the pen recorders and the ray oscilloscopes, the electrified coil is subjected to the electromagnetic torque $k_i i(t)$ in the permanent magnetic field, resulting in the deflection motion of the pointer. The moment of the deflection inertia will be affected by the torsional damping torque $C(\mathrm{d}\theta(t)/\mathrm{d}t)$ and the elastic recovery torque $k_\theta \theta(t)$. According to Newton Second Law, the relationship between the input and output of this system can be described by a second-order differential equation

$$J\frac{\mathrm{d}^2\theta(t)}{\mathrm{d}t^2} + C\frac{\mathrm{d}\theta(t)}{\mathrm{d}t} + k_\theta\theta(t) = k_i i(t) \tag{3-36}$$

In the formula, $i(t)$ is the current signal of the input moving coil; $\theta(t)$ is the angular displacement output signal of the vibrator (moving coil); $J$ is the moment of inertia of the rotating part of the vibrator; $C$ is the damping coefficient, including air damping, electromagnetic damping, oil damping, etc.; $k_\theta$ is the torsional stiffness of the hairspring; $k_i$ is the electromagnetic torque coefficient, which is related to the effective area, number of turns and magnetic induction intensity of the moving coil winding in the air gap.

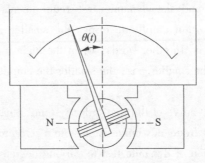

**Figure 3-12   Working principle of moving coil instrument vibrator**

After Laplace transform of Equation (3-36), the transfer function of the vibration subsystem is obtained

$$H(s)=\frac{\theta(s)}{I(s)}=\frac{\dfrac{K_i}{J}}{s^2+\dfrac{C}{J}s+\dfrac{k_\theta}{J}}=S\frac{\omega_n^2}{s^2+2\zeta_i\omega_n s+\omega_n^2} \tag{3-37}$$

In the formula, $\omega_n=\sqrt{k_\theta/J}$ is the natural frequency of the system; $\zeta=\dfrac{C}{2}\sqrt{k_\theta J}$ is the damping rate of the system; $S=k_i/k_\theta$ is the sensitivity of the system.

The frequency response characteristics of a typical second-order system are analyzed in the following content. The general formula of the differential equation of a general second-order system is

$$a_2\frac{\mathrm{d}^2 y(t)}{\mathrm{d}t^2}+a_1\frac{\mathrm{d}y(t)}{\mathrm{d}t}+a_0 y(t)=b_0 x(t) \tag{3-38}$$

After sensitivity normalization, it can be written

$$\frac{a_2}{a_0}\frac{\mathrm{d}^2 y(t)}{\mathrm{d}t^2}+\frac{a_1}{a_0}\frac{\mathrm{d}y(t)}{\mathrm{d}t}+y(t)=x(t) \tag{3-39}$$

Let $\omega_n=\sqrt{a_0/a_1}$ (called the natural frequency of the system) and $\zeta=a_1/2\sqrt{a_0 a_2}$ (called the damping rate of the system). Then

$$\frac{a_2}{a_0}=\frac{1}{\omega_n^2}$$

$$\frac{a_1}{a_0}=\frac{2\xi}{\omega_n}$$

Therefore, Equation (3-39) can be further rewritten as

$$\frac{1}{\omega_n^2}\frac{d^2y(t)}{dt^2}+\frac{2\zeta}{\omega_n}\frac{dy(t)}{dt}+y(t)=x(t) \tag{3-40}$$

Let us take the Laplace transform

$$\frac{1}{\omega_n^2}s^2Y(s)+\frac{2\zeta}{\omega_n}sY(s)+Y(s)=X(s) \tag{3-41}$$

Therefore, the transfer function of the second-order system is

$$H(s)=\frac{1}{\dfrac{1}{\omega_n^2}s^2+\dfrac{2\zeta}{\omega_n}s+1}=\frac{\omega_n^2}{s^2+2\zeta\omega_ns+\omega_n^2} \tag{3-42}$$

The frequency response of a second-order system is

$$\begin{cases} H(j\omega)=\dfrac{1}{1-\left(\dfrac{\omega}{\omega_n}\right)^2+j2\zeta\left(\dfrac{\omega}{\omega_n}\right)} \\[4mm] A(\omega)=\dfrac{1}{\sqrt{\left[1-\left(\dfrac{\omega}{\omega_n}\right)^2\right]^2+4\zeta^2\left(\dfrac{\omega}{\omega_n}\right)^2}} \\[4mm] \varphi(\omega)=-\arctan\dfrac{2\zeta\left(\dfrac{\omega}{\omega_n}\right)}{1-\left(\dfrac{\omega}{\omega_n}\right)^2} \end{cases} \tag{3-43}$$

The amplitude-frequency curve and phase-frequency curve of the second-order system are shown in Figure 3-13. It is important to note that this is the curve after sensitivity normalization. From the amplitude-frequency and phase-frequency curves of the second-order system, the main parameters that affect the system characteristics are frequency ratio $\omega/\omega_n$ and damping rate $\zeta$. Only when $\omega/\omega_n<1$ is close to the origin of the coordinate, $A(\omega)$ is closer to a horizontal straight line, and $\varphi(\omega)$ is also approximately linear with $\omega$, which can be used for dynamic non-distortion testing. If the natural frequency $w_n$ of the test system is higher, the horizontal straight line segment of $A(\omega)$ is correspondingly longer, and the operating frequency range of the system will be larger. In addition, when the damping rate $\zeta$ of the system is about 0.7, the horizontal straight section of $A(\omega)$ will be correspondingly longer, and the line between $\varphi(\omega)$ and $\omega$ is also closer to linear in a wider frequency range. When $\zeta=0.6\sim0.8$, more suitable comprehensive characteristics can be obtained. The analysis shows that when $\zeta=0.7$, within the range of $\omega/\omega_n=0\sim0.58$, the change of $A(\omega)$ does not exceed 5%, and $\varphi(\omega)$ is also close to the oblique line over the coordinate origin. It can be seen that the main dynamic performance parameters of the second-order system are the natural frequency $\omega_n$ and damping rate $\zeta$.

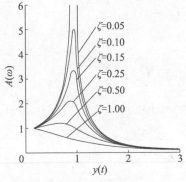
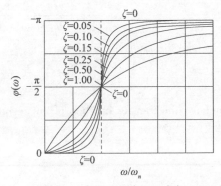

(a) Amplitude-frequency characteristic curve      (b) Phase-frequency characteristic curve

**Figure 3-13    Amplitude-frequency and phase-frequency characteristic curves of a second-order system**

Note that for the second-order system, when $\omega/\omega_n = 1$, $A(\omega) = 1/2\zeta$, if the damping rate of the system is very small, the output amplitude will increase sharply, so when $\omega/\omega_n = 1$, the system resonates. In resonance, the amplitude increases inversely proportional to the damping rate $\zeta$, and no matter how large the damping rate is, the phase of the output of the system always lags behind the input by 90°. In addition, after $\omega/\omega_n > 2.5$, $\varphi(\omega)$ is close to 180°, and $A(\omega)$ is also close to a horizontal straight segment, but the output is much smaller than the input.

Test devices such as mass-spring-damping system and RLC circuit are all second-order systems, as shown in Figure 3-14 and Figure 3-15.

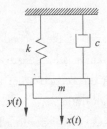
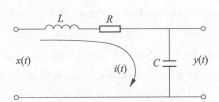

**Figure 3-14    Mass-spring system**      **Figure 3-15    RLC circuit**

### 3.3.2.3   Responses of the first and second order systems to unit-step input

The unit step signal shown in Figure 3-16.

$$x(t) = \begin{cases} 1 & (t \geq 0) \\ 0 & (t < 0) \end{cases}$$

Its Laplace transform is

$$X(s) = \frac{1}{s}.$$

The unit step response of the first-order system is shown in Figure 3-17.

$$y(t) = 1 - e^{-t/\tau} \tag{3-44}$$

The unit step response of the second-order system is shown in Figure 3-18.

$$y(t) = 1 - \frac{e^{-\xi\omega_n t}}{\sqrt{1-\zeta^2}} \sin(\omega_d t + \varphi) \tag{3-45}$$

Where $\omega_d = \omega_n \sqrt{1-\zeta^2}$ ; $\varphi = \arctan \dfrac{\sqrt{1-\zeta^2}}{\zeta} (\zeta < 1)$ .

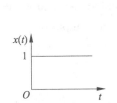

**Figure 3-16　Unit step input**

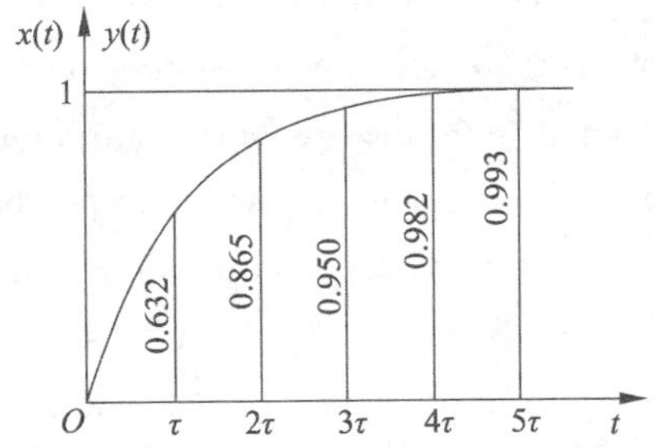

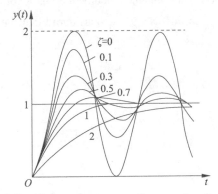

**Figure 3-17　Unit step response of a first-order system**

**Figure 3-18　Unit step response of a second-order system**

　　As can be seen from the above figure, the steady-state output error of the first-order system under unit step excitation is zero, and the transition time to steady state is $t \to \infty$. But when $t = 4\tau$, $y(4\tau) = 0.982$, the error is less than 2%; when $t = 5\tau$, $y(5\tau) = 0.993$, the error is less than 1%; so for a first-order system, the smaller the time constant $\tau$, the faster the response.

　　The steady-state output error of the second-order system under unit step excitation is also zero. The transition time to steady state depends on the natural frequency $\omega_n$ and the damping ratio $\zeta$ of the system. The higher the $\omega_n$, the faster the system response. The damping ratio mainly affects the overshoot and the number of oscillations. When $\zeta = 0$, the overshoot is 100%, and it continues to oscillate; when $\zeta \geqslant 1$, it is essentially a series of two first-order systems. Although there is no oscillation, it takes a long time to reach steady state. Usually, while $\zeta = 0.6 \sim 0.8$, the maximum overshoot at this time is no more than 10%, and the time to reach the steady state is the shortest, which is about $(5 \sim 7)/\omega_n$, and the steady state error is within the range of $2\% \sim 5\%$. Therefore, the damping ratio of the second-order test system is usually chosen as $\zeta = 0.6 \sim 0.8$.

　　In engineering application, the sudden loading or unloading of the system can be regarded as the first step input to the system. Because applying this kind of input is simple and easy, and can fully reflect the dynamic characteristics of the system, it is often used for dynamic calibration of the system.

### 3.3.3 Test conditions for non-distortion testing

Due to the influence of the actual test system, there will always be more or less distortion during the test. The so-called non-distortion test system is that when the response of the test system is compared with the excitation waveform, only the amplitude size and the time of appearance are different, other aspects have no change. If the input and output of the test system are $x(t)$ and $y(t)$ respectively, the meaning of non-distortion test can be expressed as

$$y(t) = A_0 x(t-t_0) \qquad (3\text{-}46)$$

In the formula, $t_0$ is the lag time, both $A_0$ and $t_0$ are constants.

This formula shows that the output waveform of this device is exactly the same as the input waveform, but the amplitude (or each instantaneous value) is amplified by $A_0$ times and delayed by $t_0$ in time (see Figure 3-19). In this case, it is considered that the measurement device has the characteristics of non-distortion test.

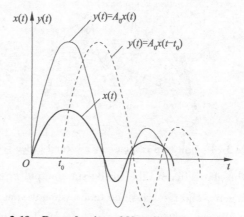

**Figure 3-19    Reproduction of Non-distortion test waveform**

Now, according to Equation (3-46) to analyze the frequency characteristics of the device to achieve test without distortion. Taking the Fourier transform of this equation, the system frequency response function can be obtained as

$$H(\omega) = A(\omega)\,e^{-j\omega t_0} \qquad (3\text{-}47)$$

If it is not to be distorted, it must be satisfied

$$\begin{cases} A(\omega) = A_0 = \text{constant} \\ \phi(\omega) = -t_0\omega \end{cases} \qquad (3\text{-}48)$$

In the formula, $A_0$ and $t_0$ are constants.

The amplitude-frequency and phase-frequency characteristic curves of an ideal non-distortion test system are shown in Figure 3-20. It can be seen that the condition for the test system to achieve non-distortion test in the frequency domain is that the amplitude-frequency characteristic curve is a straight line parallel to the $\omega$ axis, and the phase-frequency characteristic curve is a straight line with a slope of $-t_0$.

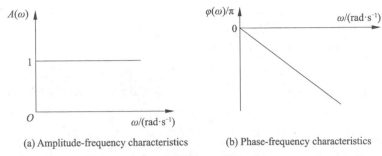

(a) Amplitude-frequency characteristics　　　(b) Phase-frequency characteristics

**Figure 3-20　Characteristics of ideal non-distortion measurement system**

In fact, the response of many linear test systems is not consistent with the excitation waveform, and most of the signals are distorted after passing through the test system. There are many reasons for this distortion. Firstly, the test system attenuates or amplifies the amplitude of each frequency component to different degrees ($A(\omega)$ is not constant), so that the relative amplitude of each frequency component in the output response changes; secondly, because the phase shift of each frequency component of the system is not proportional to the frequency (the relationship between $\varphi(\omega)$ and $x(t)$ is nonlinear), the relative position of each frequency component in the response changes; or due to the combination of the above two distortions. The distortion caused by $A(\omega)$ not being constant is called amplitude distortion, and the distortion caused by the nonlinear relationship between $\varphi(\omega)$ and $\omega$ is called phase distortion.

The actual test device cannot meet the requirements of the non-distortion test condition in a very wide frequency range, so the test device usually has both amplitude distortion and phase distortion. Figure 3-21 shows four signals with different frequencies after they are input into the test device. The amplitude-frequency characteristic of this device can be expressed as $A(\omega)$, and the phase-frequency characteristic can be expressed as $\varphi(\omega)$. The four input signals are sinusoidal signals (including the DC signal), and at a certain reference time, $t=0$, all initial phase angles are zero. As shown in Figure 3-21, it can be seen that the output signal has different amplitude gain and phase lag compared with the input signal. For the signal with single frequency component, because the linear system usually has frequency retention, as long as its amplitude is not in the nonlinear region, the output signal frequency is also single, so there is no distortion.

For those with multiple frequency components, it is obvious that both amplitude distortion and phase distortion are excited. Especially in the transition stage of the frequency component before and after $\omega_n$, the signal distortion is particularly serious.

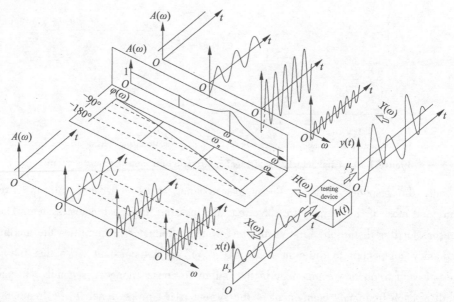

**Figure 3-21   Output of different frequency components in the signal after passing the test device**

It should be noted that if the purpose of the test is to accurately obtain the signal waveform, the non-distortion condition expressed by Equation ( 3-48 ) fully meets the requirements. However, the above-mentioned conditions are not complete when the signal is obtained for feedback control, because the time lag may destroy the stability of the control system, and the ideal condition $\varphi(\omega) = 0$ is still needed in this case.

In the actual test process, in order to reduce the test error caused by system distortion, in addition to selecting the suitable test system according to the frequency band of the signal under test, the input signal is usually preprocessed to reduce or eliminate interference signals and improve the signal-to-noise ratio as much as possible. In addition, when selecting and designing a test system, the test parameters should be selected reasonably according to the signal to be tested.

For example, in vibration testing or fault diagnosis, it is often only necessary to test the frequency components and their amplitude in the vibration, instead of studying its changing waveforms. In this case, the amplitude-frequency characteristic or amplitude-distortion is the most important index, while the phase-frequency characteristic or phase-distortion index does not have strict precision requirements. For another example, when we need to obtain the delay time of a specific waveform through testing, there are strict requirements for the phase-frequency characteristics of the testing device, so as to reduce the test error caused by phase distortion.

To make the test system accurate and reliable, the calibration of the test system should not only be accurate, but also should be calibrated regularly. In essence, the calibration is the measurement of the characteristic parameters of the measuring device itself.

When measuring the static parameters of the device, the calibrated " standard " static quantity is generally taken as the input, so that it can determine the input-output characteristic curve. Based on this curve, the return error, calibration curve, linear error and sensitivity can be further determined. The input error used should be no more than 1/3 ~ 1/5 or less of the predetermined test error.

# Questions

3.1 Explain the role of linear system frequency retention in measurement.

3.2 When using a piezoelectric force sensor with a sensitivity of 80 nC/MPa for pressure measurement, first connect it to a charge amplifier with a gain of 5 mV/mC, and the charge amplifier is connected to a written test recorder with a sensitivity of 25 mm/V, to try to find the sensitivity of the pressure test system. What is the pressure change when the output of the recorder changes by 30 mm?

3.3 Connect the piezoelectric force sensor with sensitivity $4.04 \times 10^{-2}$ pC/Pa to a charge amplifier whose sensitivity is adjusted to 0.226 mV/pC, and find its total sensitivity. If you want to adjust the total sensitivity to $10^7$ mV/Pa, how to adjust the sensitivity of the charge amplifier.

3.4 Briefly describe the undistorted test system and its satisfying conditions.

3.5 A first-order device with a time constant of 0.35 s is used to measure sinusoidal signals with periods of 1 s, 2 s, and 5 s, and what is the amplitude error of the steady-state response?

3.6 If you want to use a first-order system to measure a 100 Hz sinusoidal signal, and you want to limit the amplitude error to 5%, what is the time constant? If the system is used to measure a 50 Hz sinusoidal signal, what is the amplitude error and phase angle difference at this time?

3.7 Calculate the steady-state response of the periodic signal $x(t) = 0.5\cos 10t + 0.2\cos (100t - 45°)$ through the device with the transfer function $H(s) = 1/(0.005s+1)$.

3.8 Why is the damping ratio $\zeta$ of the second-order device in the range of 0.6 to 0.8?

3.9 Suppose a force sensor can be treated as a second-order oscillation system. It is known that the natural frequency of the sensor is 800 Hz, and the damping ratio $\zeta = 0.14$. When using this sensor for a sine force test with a frequency of 400 Hz, what is the amplitude ratio $A(\omega)$ and the phase angle difference $\varphi(\omega)$? If the damping ratio of the device is changed to $\zeta = 0.7$, how will $A(\omega)$ and $\varphi(\omega)$ change?

3.10 After inputting a unit step function to a device that can be regarded as a second-order system, the first overshoot peak value of the measured response is 1.15, and the oscillation period is 6.28 s. Assuming that the static gain of the device is 3, find the transfer function of the device and the frequency response of the device at the undamped natural frequency.

# Chapter 4

## Signal Description Method

In engineering and scientific research, it is often necessary to observe many objects or physical processes to obtain information about the state and movement of the research object. The information of the researched object is often very rich, and the test work is to obtain interesting and limited specific information in the signal according to certain purposes and requirements, but not all the information. In order to achieve the testing purpose, it is necessary to study various description methods of signals. This chapter introduces the basic time domain and frequency domain description methods of signals.

## 4.1  Signal classification

According to mathematical relationships, value characteristics, energy power, signals can be divided into deterministic signals and non-deterministic signals, continuous signals and discrete signals, energy signals and power signals, etc.

### 4.1.1  Deterministic signal and random signal

#### 4.1.1.1  Deterministic signal

Deterministic signals can be divided into periodic signals and aperiodic signals. Periodic signals can be divided into sinusoidal periodic signals and complex periodic signals. Aperiodic signals can be further divided into quasi-periodic signals and transient signals. The classification of deterministic signals is shown in Figure 4-1.

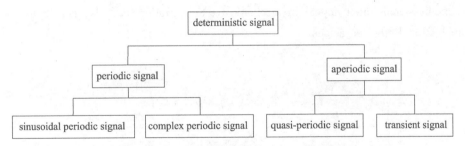

**Figure 4-1   Classification of deterministic signals**

(1) Periodic signal

The signal that appears repeatedly after a period of time interval is called periodic signal, and the most basic periodic signal is sinusoidal signal, which can be expressed as:

$$x(t) = A\sin(2\pi ft + \theta_0) \tag{4-1}$$

Where $A$ is the amplitude, $f$ is the vibration frequency, and $\theta_0$ is the initial phase angle.

The complex periodic signal is composed of the superposition of sinusoidal signals of different frequencies, and the ratio of the frequencies is a rational number. If the fundamental frequency in the periodic signal is $f$, then the frequency $f_n$ of each sinusoidal signal is an integer multiple of the fundamental frequency, namely $f_n = nf(n = 1, 2, \cdots)$.

In the mechanical system, the vibration caused by the unbalance of the rotating body is often a periodic motion. Figure 4-2 is the vibration signal (Measurement point 3) measured on a certain reducer, which can be approximated as a periodic complex signal.

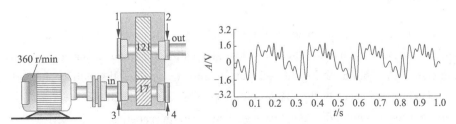

**Figure 4-2   Vibration measuring point layout and vibration signal of a reducer**

(2) Aperiodic signal

A signal that can be described by a clear mathematical relationship but does not have periodic repetition is called an aperiodic signal. It is divided into quasi-periodic signals and transient signals. The quasi-periodic signal is formed by the superposition of two or more sinusoidal signals with different frequencies, but the frequency ratio is not all rational number.

In actual machinery, when several different periodic vibration sources interact with each other, quasi-periodic signals are often generated. For example, the machine tool vibration caused by the asynchronous vibration of several motors, and the measured results of the signals are quasi-periodic signals. For another example, $x(t) = \sin t + \sin\sqrt{2}\,t$ is the synthesis of two sinusoidal signals, but the frequency ratio is not a rational number, it is a quasi-periodic signal. The signal waveform is shown in Figure 4-3. It cannot be seen from the waveform that it is a periodic signal.

Quasi-periodic signals often appear in mechanical rotor vibration analysis, gear noise analysis, voice analysis and other occasions.

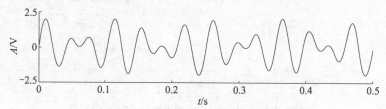

Figure 4-3    Quasi-periodic signal

Aperiodic signals other than quasi-periodic signals are transient signals, which appear at a certain moment and disappear at a certain moment. There are many factors that produce transient signals, such as the free oscillation of the damped oscillation system after the excitation force is removed. Figure 4-4 shows the response of a single-degree-of-freedom vibration model under impulse force, which is a transient signal.

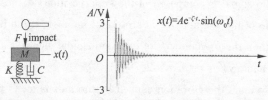

Figure 4-4    Single-degree-of-freedom vibration model impulse response signal

### 4.1.1.2    Random signal

The signal that can not be accurately predicted the future instantaneous value, and can not be described by accurate mathematical relations, is called the random signal, also called the uncertain signal. As shown in Figure 4-5, the vibration signal of the lathe spindle under the influence of the environment in the machining process obviously cannot be accurately predicted a certain instantaneous amplitude, but this signal has certain statistical characteristics. When the number of experiments is large or the signal is obtained for a long time, the average value of its amplitude may tend to a certain limit value.

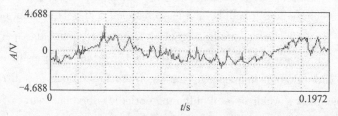

Figure 4-5    Vibration signal of lathe spindle under the influence of the
environment in the machining process

## 4.1.2    Continuous signal and discrete signal

If the value of the independent variable in the signal mathematical expression is continuous, it is called a continuous signal, as shown in Figure 4-6a. If the independent variables take discrete values, it is called a discrete signal. Figure 4-6b is the result of continuous signal sampling at equal time interval, which is a discrete signal. Discrete signals can be represented by discrete graphs or by numerical sequences. The amplitude of the signal is also classified as continuous and discrete. If the amplitude and independent variable of the signal are continuous, it is called analog signal. If the signal amplitude and independent variables are discrete, it is called a digital signal. At present, all the signals used in digital computers are digital signals.

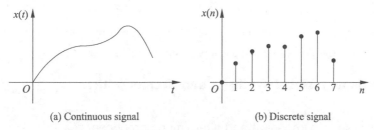

(a) Continuous signal　　　　(b) Discrete signal

**Figure 4-6　Continuous signal and discrete signal**

### 4.1.3　Energy signal and power signal

In non-electric quantity measurement, the measured signal is usually converted into a voltage or current signal for processing. The voltage signal $x(t)$ is applied to the resistor $R$ and its instantaneous power is $P(t) = x^2(t)/R$. When $R = 1$, then $P(t) = x^2(t)$. The integral of instantaneous power over time is the energy of the signal in that time. Usually, the actual unit of the signal is ignored, and the square $x^2(t)$ of the signal $x(t)$ and its integral over time are respectively called the power and energy of the signal. While the following condition is satisfied

$$\int_{-\infty}^{+\infty} x^2(t)\ \mathrm{d}t < \infty \qquad (4\text{-}2)$$

Then, the energy of a signal is considered to be limited, and it is called a limited energy signal, or energy signal for short, such as rectangular pulse signal, attenuation index signal, etc.

If the signal is in the interval $(-\infty, +\infty)$, the energy is infinite.

$$\int_{-\infty}^{+\infty} x^2(t)\,\mathrm{d}t \to \infty \qquad (4\text{-}3)$$

But its average power in the finite interval $(t_1, t_2)$ is limited, that is

$$\frac{1}{t_2 - t_1}\int_{t_1}^{t_2} x^2(t)\ \mathrm{d}t < \infty \qquad (4\text{-}4)$$

Such signals are called power limited signals, or power signals, such as various periodic signals, step signals, etc. It must be noted that the power and energy of the signal may not have the unit of true power and energy.

# 4.2 Time domain description of the signal

Dynamic signal is usually a physical quantity that varies with time, which is called the time domain waveform of the signal. In order to understand the properties of signals from the time domain waveform, the complex signal can be decomposed into several simple signals from different perspectives, or the evaluation of operation state for the tested object can be obtained directly through the time domain statistical characteristic parameters.

## 4.2.1 Time domain signal synthesis and decomposition

### 4.2.1.1 Steady-state component and alternating component

Generally, the signal $x(t)$ can be decomposed into the sum of the steady-state component $x_d(t)$ and alternating component $x_a(t)$, as shown in Figure 4-7.

$$x(t) = x_d(t) + x_a(t) \tag{4-5}$$

The steady-state component is a quantity that varies regularly and is sometimes called a trend quantity. The alternating component may contain the amplitude, frequency, and phase information of the physical process under study, or it may be random interference noise.

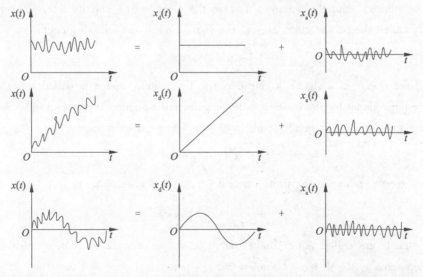

**Figure 4-7 The signal is decomposed into the sum of steady-state components and alternating components**

### 4.2.1.2 Even component and odd component

The signal $x(t)$ also can be decomposed into the sum of even component $x_e(t)$ and odd component $x_o(t)$, as shown in Figure 4-8. The even component is symmetric about the longitudinal axis, and the odd component is symmetric about the origin.

$$x(t) = x_e(t) + x_o(t) \tag{4-6}$$

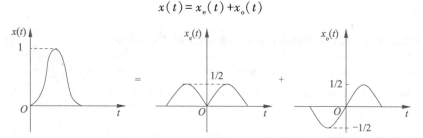

**Figure 4-8　The signal is decomposed into the sum of odd and even components**

### 4.2.1.3　Real part and imaginary part

For the signal $x(t)$ whose instantaneous value is a complex number can be decomposed into the sum of real and imaginary parts, i.e

$$x(t) = x_R(t) + jx_I(t) \tag{4-7}$$

Generally, the actual signals generated are mostly real signals, but in signal analysis, some real signals are often studied with the help of the complex number form of the signal, because this analysis method can simplify the calculation. For example, regarding the measurement of axis rotation accuracy and its signal processing, the error movement of the rotation axis along the radius direction is usually regarded as the periodic movement of a point on the plane. It can be represented by a complex number with time as the independent variable $x(t)$. The real part $x_R(t)$ and imaginary part $x_I(t)$ can be measured with mutually perpendicular radial measuring devices, and the signal $x(t)$ obtained is the sum of the two.

### 4.2.1.4　Orthogonal function component

The signal $x(t)$ can be represented by a set of orthogonal functions, i.e

$$x(t) \approx c_1 x_1(t) + c_2 x_2(t) + \cdots + c_n x_n(t) \tag{4-8}$$

The orthogonal condition for each component is

$$\int_{t_1}^{t_2} x_i(t) x_j(t)\, dt = \begin{cases} 0 & (i \neq j) \\ k & (i = j) \end{cases} \tag{4-9}$$

That is, the integral of the product of different components in the interval $(t_1, t_2)$ is zero, and the energy of any component in the interval $(t_1, t_2)$ is finite. The coefficient $x(t) = y(t)$ of each component in Equation (4-8) is obtained from the following equation under the condition of satisfying the minimum mean square error.

$$c_i = \frac{\displaystyle\int_{t_1}^{t_2} x(t) x_i(t)\, dt}{\displaystyle\int_{t_1}^{t_2} x_i^2(t)\, dt} \tag{4-10}$$

The set of functions that satisfy the orthogonality condition include trigonometric functions, complex exponential functions, etc. For example, when a signal is described by a set of trigonometric functions, the signal $x(t)$ can be decomposed into the sum of many positive (cosine) sine trigonometric functions.

## 4.2.2 Statistical characteristic parameters of the signal

Some statistical characteristic parameters that can be obtained directly through time-domain waveforms are often used to quickly evaluate or diagnose the state of mechanical systems.

(1) Mean value

The mean value is the average of the sample function $x(t)$ of the random signal over the entire time coordinate. Namely

$$\mu_x = \lim_{T \to \infty} \frac{1}{T} \int_0^T x(t)\,dt \tag{4-11}$$

In actual processing, since infinite time sampling is impossible, only finite length samples can be used for estimation.

$$\hat{\mu}_x = \frac{1}{T} \int_0^T x(t)\,dt \tag{4-12}$$

The physical meaning of the mean value represents the magnitude of the DC component in the signal and describes the static component of the signal.

(2) Mean square value

The mean square value is the mean of the signal squared value, which is expressed as

$$\psi_x^2 = \lim_{T \to \infty} \frac{1}{T} \int_0^T x^2(t)\,dt \tag{4-13}$$

The mean square value is calculated to be

$$\hat{\psi}_x^2 = \frac{1}{T} \int_0^T x^2(t)\,dt \tag{4-14}$$

Its physical meaning indicates the strength or power of the signal. The positive square root of the mean square value is called the root mean square value $\hat{x}_{rms}$, also known as the effective value.

$$\hat{x}_{rms} = \sqrt{\hat{\psi}_x^2} = \sqrt{\frac{1}{T} \int_0^T x^2(t)\,dt} \tag{4-15}$$

It is another expression of signal average energy (or power).

(3) Variance

The variance of the signal $x(t)$ describes the fluctuation degree of the random signal amplitude, which is defined as

$$\sigma_x^2 = \lim_{T \to \infty} \frac{1}{T} \int_0^T [x(t) - \mu_x]^2\,dt \tag{4-16}$$

The square root of variance $\sigma_x$ describes the dynamic component of the signal.

The relationship among the mean $\mu_x$, the mean square $\psi_x^2$ and the variance $\sigma_x^2$ is as follows.

$$\psi_x^2 = \mu_x^2 + \sigma_x^2 \tag{4-17}$$

## 4.2.3   Application of statistical characteristic parameters

### 4.2.3.1   Diagnostic method by using root mean square value

It is the simplest and most commonly used method to use the root mean square value of the vibration response of some characteristic points on the mechanical system as the basis for judging the fault. For example, in China, it has stipulated that the vertical vibration displacement amplitude on the bearing seat of turbine generator shall not exceed 0.05 mm, if not, it should be shut down for maintenance.

The root-mean-square diagnosis method can be applied to equipment that performs simple harmonic vibration, equipment that performs periodic vibration, and can also be used for equipment that performs random vibration. Generally, in terms of vibration measurement, displacement should be measured at low frequencies (tens of Hz); velocity should be measured at intermediate frequencies (about 1000 Hz); acceleration should be measured at high frequencies. The permissible vibration levels of rotary machinery specified in ISO 2372 and ISO 2373 of the International Standards Institute are shown in Table 4-1.

**Table 4-1   The permissible vibration level of rotating machinery**           mm/s

| Limit | Normal limit | High limit | Warning limit | Parking limit |
|---|---|---|---|---|
| Small machinery | 0.28~0.71 | 1.80 | 4.50 | 7.10~71.0 |
| Medium machinery | 0.28~1.12 | 2.80 | 7.10 | 11.2~71.0 |
| Large machinery | 0.28~1.80 | 4.50 | 11.2 | 18.0~71.0 |
| Extra large machinery | 0.28~2.80 | 7.10 | 18.0 | 28.0~71.0 |

### 4.2.3.2   Diagram diagnostic method based on amplitude-time

The root-mean-square value diagnosis method is most suitable for the steady-state vibration of the machine.

The amplitude-time graph diagnosis method is mostly used to measure and record the variation of the vibration amplitude with time in the process of starting and stopping the machine, and to judge the fault of the machine according to the amplitude time curve. Take the starting process of centrifugal air compressor or other rotating machinery as an example, the recorded amplitude changes with time are shown in Figure 4-9.

Figure 4-9a shows that the amplitude does not change with the start-up process. It may be caused by other equipment and foundation vibration transmitted to the tested equipment, or it may be caused by fluid pressure pulsation or valve vibration.

Figure 4-9b shows that the amplitude increases with the start-up process, which may be caused by poor dynamic balance of the rotor, low stiffness of the bearing seat, or damage to the thrust bearing, etc.

Figure 4-9c shows the peak amplitude during the boot process, which is mostly caused by resonance. Including the so-called flexible rotors where the critical speed of the shaft system is lower than the working speed, as well as the resonance of the box, support, and foundation.

Figure 4-9d shows that the amplitude suddenly increases at some point during the start-up process, which may be caused by oil film vibration or may be caused by too small clearance or insufficient interference.

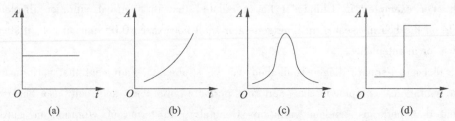

**Figure 4-9    Amplitude time chart of the boot process**

It should be noted that: large rotating machinery often uses dynamic pressure bearings to support the rotor. The bearing relies on the oil film to achieve the cooperation of dynamic and static parts. When the oil film thickness, pressure, viscosity, temperature and other parameters are constant, the rotor may vibrate suddenly after reaching a certain speed. When the speed increases again, the amplitude does not decrease, this is the oil film vibration.

If the gap is too small, when the deformation caused by temperature or centrifugal force reaches a certain value, it will cause a collision, causing the amplitude to suddenly increase. Another example is the insufficient interference between the impeller of the blade machinery and the sleeve of the rotating shaft, when the centrifugal force reaches a certain value, it will cause looseness, and the amplitude will also increase suddenly.

# 4.3    Frequency-domain description of the signal

The time-domain description for a signal takes time as an independent variable, reflecting the relationship of signal amplitude with time. In practice, the signal is more complicated, and time-domain analysis cannot completely extract all the information. Therefore, in order to study the signal more comprehensively and deeply and obtain more useful information, the time-domain description is often transformed into the frequency-domain description, that is, the frequency is used as an independent variable to represent the signal. The time and frequency-domain descriptions for signals can be mutually converted, and they contain the same amount of information. This section will describe and analyze periodic signals and non-periodic signals from both time domain and frequency domain.

## 4.3.1　Description of periodic signal

The harmonic signal is the simplest periodic signal with only one frequency component. Generally, a periodic signal can be expanded into a linear superposition of multiple or even an infinite number of harmonic signals with different frequencies by using Fourier series.

### 4.3.1.1　Expansion of trigonometric function of periodic signal

If the periodic signal $x(t)$ satisfies the Dirichlet condition, that is, there are continuous or only a finite number of discontinuous points of the first type in the period interval $(-T_0/2, T_0/2)$, and only a finite number of extreme value points, then $x(t)$ can be expanded as

$$x(t) = a_0 + \sum_{n=1}^{+\infty} (a_n \cos n\omega_0 t + b_n \sin n\omega_0 t) \tag{4-18}$$

Where the constant value component $a_0$, the cosine component amplitude $a_n$, and the sine component amplitude $b_n$ are as follows.

$$a_0 = \frac{1}{T_0} \int_{-\frac{T_0}{2}}^{\frac{T_0}{2}} x(t)\, dt$$

$$a_n = \frac{2}{T_0} \int_{-\frac{T_0}{2}}^{\frac{T_0}{2}} x(t) \cos n\omega_0 t\, dt$$

$$b_n = \frac{2}{T_0} \int_{-\frac{T_0}{2}}^{\frac{T_0}{2}} x(t) \sin n\omega_0 t\, dt \tag{4-19}$$

$$\omega_0 = \frac{2\pi}{T_0}$$

Where $a_0, a_n, b_n$ is the Fourier coefficient; $T_0$ is the period of the signal; $\omega_0$ is called the fundamental angular frequency; $n\omega_0$ is the $n$ subharmonic frequency.

By the trigonometric function transformation, Equation (4-19) can be rewritten as

$$x(t) = A_0 + \sum_{n=1}^{+\infty} A_n \sin(n\omega_0 t + \varphi_n) \tag{4-20}$$

Where the constant value component $A_0$, the amplitude $A_n$ of each harmonic component, and the initial phase angle $\psi_n$ of each harmonic component are as follows.

$$A_0 = a_0$$

$$A_n = \sqrt{a_n^2 + b_n^2}$$

$$\varphi_n = \arctan \frac{a_n}{b_n} \tag{4-21}$$

Equation (4-21) shows that any periodic signal satisfying Dirichlet condition can be decomposed into a constant component and several different harmonic signal components, and the angular frequency of these harmonic signal components is an integer multiple of the fundamental wave angular frequency.

Taking frequency $\omega(n\omega_0)$ as the abscissa, and taking the amplitude $A_n$ and phase $\varphi_n$ as the ordinate respectively, then, $A_n-\omega$ is called the amplitude spectrogram of the signal, and $\varphi_n-\omega$ is called the phase spectrogram. The two are collectively called the frequency spectrum of the signal. The frequency components, amplitudes and phases of periodic signals can be clearly and intuitively seen from the spectrogram.

**Example 4-1**  Find the Fourier series of the periodic square wave in Figure 4-10.

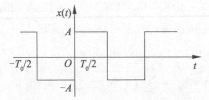

**Figure 4-10  Periodic square wave**

**Solution**: $x(t)$ in a cycle of $[-T_0/2, T_0/2]$, the expression is

$$x(t) = \begin{cases} A & \left(0 < t < \dfrac{T_0}{2}\right) \\[3mm] -A & \left(-\dfrac{T_0}{2} < t \leqslant 0\right) \end{cases}$$

Since $x(t)$ is an odd function, and the integral value of the odd function in one cycle is 0.

$$a_0 = \frac{1}{T_0} \int_{-\frac{T_0}{2}}^{\frac{T_0}{2}} x(t)\,\mathrm{d}t = 0, a_n = 0$$

$$b_n = \frac{2}{T_0} \int_{-\frac{T_0}{2}}^{\frac{T_0}{2}} x(t)\sin n\omega_0 t\,\mathrm{d}t$$

$$= \frac{2}{T_0}\left(\int_{-\frac{T_0}{2}}^{0} -A\sin n\omega_0 t\,\mathrm{d}t + \int_{0}^{\frac{T_0}{2}} A\sin n\omega_0 t\,\mathrm{d}t\right)$$

$$= \frac{2A}{T_0}\left[\left(\frac{\cos n\omega_0 t}{n\omega_0}\right)\Big|_{-\frac{T_0}{2}}^{0} + \left(-\frac{\cos n\omega_0 t}{n\omega_0}\right)\Big|_{0}^{\frac{T_0}{2}}\right]$$

$$= \frac{2A}{n\omega_0 T}\left[1 - \cos\left(-\frac{n\omega_0 T_0}{2}\right) - \cos\left(\frac{n\omega_0 T_0}{2}\right) + 1\right]$$

$$= \frac{4A}{n\omega_0 T}\left[1 - \cos\left(\frac{n\omega_0 T_0}{2}\right)\right]$$

$$= \begin{cases} \dfrac{4A}{n\pi} & (n = 1,3,5,\cdots) \\[3mm] 0 & (n = 2,4,6,\cdots) \end{cases}$$

Therefore, the following result is obtained.

$$x(t) = \frac{4A}{\pi}\left(\sin \omega_0 t + \frac{1}{3}\sin 3\omega_0 t + \frac{1}{5}\sin 5\omega_0 t + \cdots\right)$$

According to the above analysis, the amplitude spectrum and phase spectrum are shown in Figures 4-11a and 4-11b respectively. The amplitude spectrum only contains the frequency components of the fundamental and odd harmonics, and the initial phase $\varphi_n$ of each harmonic component in the phase spectrum is zero.

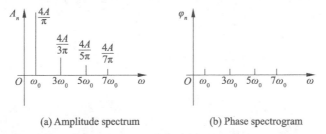

(a) Amplitude spectrum　　　　(b) Phase spectrogram

**Figure 4-11　Amplitude spectrum and phase spectrum of a periodic square wave**

Figure 4-12 shows the relationship between time domain and frequency spectrum of periodic square wave. The time domain description (waveform) and frequency domain description (spectrum) of periodic square wave and their mutual relations are illustrated vividly by means of waveform decomposition.

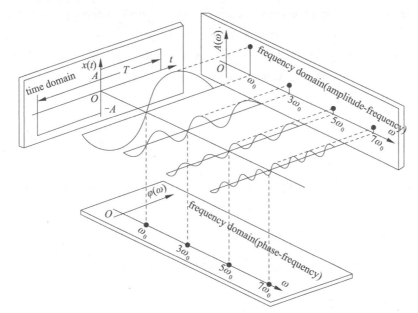

**Figure 4-12　Time domain and frequency domain description of a periodic square wave**

## 4.3.1.2　Complex exponential expansion of periodic signals

Using Euler's formula

$$e^{\pm jn\omega_0 t} = \cos n\omega_0 t \pm j\sin n\omega_0 t \tag{4-22}$$

the following equation can be obtained.

$$\cos n\omega_0 t = \frac{1}{2}(e^{-jn\omega_0 t} + e^{jn\omega_0 t}) \tag{4-23}$$

$$\sin n\omega_0 t = \frac{j}{2}(e^{-jn\omega_0 t} - e^{jn\omega_0 t}) \qquad (4\text{-}24)$$

Where $j = \sqrt{-1}$. Rewrite Equation (4-18) as the following expression.

$$x(t) = a_0 + \sum_{n=1}^{\infty}\left[\frac{1}{2}(a_n + jb_n)e^{-jn\omega_0 t} + \frac{1}{2}(a_n - jb_n)e^{jn\omega_0 t}\right] \qquad (4\text{-}25)$$

$$C_0 = a_0$$

$$C_{-n} = \frac{1}{2}(a_n + jb_n)$$

$$C_n = \frac{1}{2}(a_n - jb_n) \qquad (4\text{-}26)$$

$$x(t) = C_0 + \sum_{n=1}^{\infty}(C_{-n}e^{-jn\omega_0 t} + C_n e^{jn\omega_0 t})$$

$$x(t) = \sum_{n=-\infty}^{+\infty} C_n e^{jn\omega_0 t} \qquad (n = 0, \pm 1, \pm 2, \cdots) \qquad (4\text{-}27)$$

$$C_n = \frac{1}{T_0}\int_{-\frac{T_0}{2}}^{\frac{T_0}{2}} x(t)e^{-jn\omega_0 t}\,\mathrm{d}t \qquad (n = 0, \pm 1, \pm 2, \cdots) \qquad (4\text{-}28)$$

In general, $C_n$ is a complex number and can be written as the following expression.

$$C_n = \mathrm{Re}\ C_n + j\mathrm{Im}\ C_n = |C_n|e^{j\varphi_n} \qquad (4\text{-}29)$$

Where $\mathrm{Re}\ C_n$ and $\mathrm{Im}\ C_n$ are called the real frequency spectrum and the imaginary spectrum respectively; and $|C_n|, \psi_n$ is called the amplitude-frequency spectrum and the phase frequency spectrum respectively. The relationship between the two forms is as follows.

$$|C_n| = \sqrt{(\mathrm{Re}\ C_n)^2 + (\mathrm{Im}\ C_n)^2} \qquad (4\text{-}30)$$

$$\psi_n = \arctan\frac{\mathrm{Im}\ C_n}{\mathrm{Re}\ C_n} \qquad (4\text{-}31)$$

**Example 4-2** Try to find the complex exponential expansion of the periodic square wave shown in Figure 4-10 and make a spectrum graph.

**Solution:**

$$C_n = \frac{1}{T_0}\int_{-\frac{T_0}{2}}^{\frac{T_0}{2}} x(t)e^{-jn\omega_0 t}\,\mathrm{d}t$$

$$= \frac{1}{T_0}\int_{-\frac{T_0}{2}}^{\frac{T_0}{2}} x(t)(\cos n\omega_0 t - j\sin n\omega_0 t)\,\mathrm{d}t$$

$$= \begin{cases} -j\dfrac{2A}{n\pi} & (n = \pm 1, \pm 3, \pm 5\cdots) \\ 0 & (n = 0, \pm 2, \pm 4, \pm 6\cdots) \end{cases}$$

*Then*

$$x(t) = \sum_{n=-\infty}^{+\infty} C_n e^{jn\omega_0 t} = -j\frac{2A}{\pi}\sum_{n=-\infty}^{+\infty}\frac{1}{n}e^{jn\omega_0 t} \qquad (n = \pm 1, \pm 3, \pm 5, \cdots)$$

Amplitude spectrum

$$|C_n| = \begin{cases} \left| \dfrac{2A}{n\pi} \right| & (n = \pm 1, \pm 3, \pm 5 \cdots) \\ 0 & (n = 0, \pm 2, \pm 4, \pm 6 \cdots) \end{cases}$$

Phase spectrum

$$\varphi_n = \arctan \dfrac{-\dfrac{2A}{n\pi}}{0} = \begin{cases} -\dfrac{\pi}{2} & (n > 0) \\ \dfrac{\pi}{2} & (n < 0) \end{cases}$$

Real and imaginary spectrum

$$\begin{cases} \mathrm{Re}\ C_n = 0 \\ \mathrm{Im}\ C_n = -\dfrac{2A}{n\pi} \end{cases}$$

The real and imaginary frequency spectrum, and the amplitude and phase-frequency spectrum are shown in Figure 4-13.

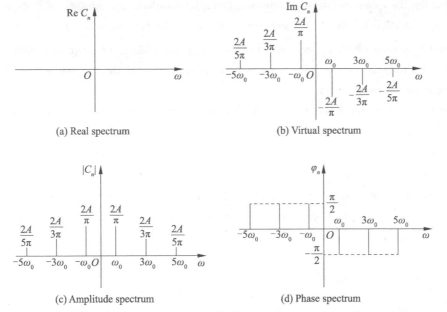

(a) Real spectrum

(b) Virtual spectrum

(c) Amplitude spectrum

(d) Phase spectrum

**Figure 4-13　Real and imaginary spectrum, amplitude and phase spectrum of a periodic square wave**

Comparing Figure 4-11 with Figure 4-13, it can be found that, in Figure 4-11, each spectral line represents the amplitude of a component, while in Figure 4-13, the amplitude of each component is divided into two parts, each half at the position corresponding to the positive and negative frequencies. The amplitude of a component can only be represented by adding the two spectral line vectors corresponding to the positive and negative frequencies. It should be noted that the occurrence of the negative frequency term is completely the result of mathematical calculation and does not have any physical significance.

From the above analysis, it can be seen that the periodic signal spectrum, whether it is a trigonometric function expansion or a complex exponential expansion, has the following characteristics:

① Discreteness

The spectrum of a periodic signal is a discrete spectrum, and each spectral line represents a sine component.

② Harmonic

The frequency of the periodic signal is composed of integer multiples of the fundamental frequency.

③ Convergence

For periodic signals that satisfy Dirichlet conditions, the general trend of harmonic amplitude is to decrease with the increase of harmonic frequency. Due to the convergence of periodic signals, it is not necessary to take harmonic components with too high order in engineering measurement.

### 4.3.2 Description of aperiodic signal

From the viewpoint of signal synthesis, multiple harmonic components with its frequency ratio being a rational number, is a periodic signal after the superposition. When the frequency ratio is not a rational number, such as $x(t) = \cos \omega_0 t + \cos \sqrt{3} \omega_0 t$, its frequency ratio is $1/\sqrt{3}$, which is not a rational number, and there is no common divisor of the frequency after synthesis, and there is no common period.

Due to the spectrum of this kind of signal is still discrete (there are two spectral lines at $\omega_0$ and $\sqrt{3} \omega_0$ respectively), it is called a quasi-periodic signal. In engineering practice, quasi-periodic signals are still very common, such as the vibration response when two or more unrelated vibration sources excite the same measured object under test. Besides, non-periodic signals generally refer to transient signals. Figure 4-14 shows an example of a transient signal. Its characteristic is that the function decays along the independent variable $t$, so the integral for this signal has a finite value, which is an energy-finite signal.

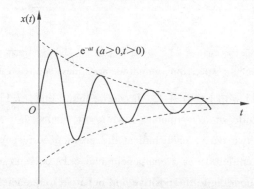

**Figure 4-14    Examples of transient signals**

### 4.3.2.1   Fourier transform

The aperiodic signal can be regarded as the periodic signal that the cycle $T_0$ goes to infinity.

When the cycle $T_0$ increases, the interval $(-T_0/2, T_0/2)$ tends to $(-\infty, \infty)$, the frequency interval of the spectrum is $\Delta\omega = \omega_0 = 2\pi/T_0 \rightarrow d\omega$, the discrete frequency $n\omega_0$ becomes continuous frequency $\omega$, and the superposition relationship of the equation expansion becomes the integral relationship, then the Equation (3-27) can be rewritten as

$$\lim_{T_0 \to \infty} x(t) = \lim_{T_0 \to \infty} \sum_{n=-\infty}^{+\infty} C_n e^{jn\omega_0}$$

$$= \lim_{T_0 \to \infty} \frac{1}{T_0} \sum_{n=-\infty}^{+\infty} \left[ \int_{-\frac{T_0}{2}}^{\frac{T_0}{2}} x(t) e^{-jn\omega_0 t} dt \right] e^{jn\omega_0 t}$$

$$= \int_{-\infty}^{\infty} \frac{d\omega}{2\pi} \left[ \int_{-\infty}^{\infty} x(t) e^{-jn\omega t} dt \right] e^{jn\omega t}$$

$$= \frac{1}{2\pi} \int_{-\infty}^{\infty} \left[ \int_{-\infty}^{\infty} x(t) e^{-jn\omega t} dt \right] e^{jn\omega t} d\omega$$

(4-32)

Mathematically, Equation (4-32) is called the Fourier integral. Strictly speaking, the Fourier integral of an aperiodic signal $x(t)$ exists only if $x(t)$ satisfies the Dirichlet condition on a finite interval and is absolutely integrable.

For Equation (4-32), after the integration of time $t$, it is only a function of angular frequency $\omega$, denoted as $X(\omega)$, where

$$X(\omega) = \int_{-\infty}^{\infty} x(t) e^{-j\omega t} dt$$

(4-33)

$$x(t) = \frac{1}{2\pi} \int_{-\infty}^{\infty} X(\omega) e^{j\omega t} d\omega$$

(4-34)

$X(\omega)$ expressed in Equation (4-33) is called the Fourier Transform (FT) of $x(t)$, and $x(t)$ in Equation (4-34) is called the Inverse Fourier Transform (IFT) of $X(\omega)$, the two are a Fourier transform pair.

After $\omega = 2\pi f$ is substituted into Equations (4-33) and (4-34), the two equations can be written as

$$X(f) = \int_{-\infty}^{\infty} x(t) e^{-j2\pi ft} dt$$

(4-35)

$$x(t) = \int_{-\infty}^{\infty} X(f) e^{j2\pi ft} df$$

(4-36)

This can avoid the constant factor of $1/2\pi$ in the Fourier transform and simplify the calculation. The relationship is as follows.

$$X(f) = 2\pi X(\omega)$$

(4-37)

Generally, $X(f)$ is a complex function of frequency $f$, which can be written as

$$X(f) = |X(f)| e^{j\varphi(f)}$$

(4-38)

Where $|X(f)|$ is the continuous amplitude spectrum of the signal $x(t)$, and $\varphi(f)$ is the continuous phase spectrum of the signal $x(t)$.

It should be noted that although the amplitude spectrum $|X(f)|$ of the aperiodic signal is very similar to the amplitude spectrum $|C_n|$ of the periodic signal, there are differences between them. The difference is highlighted by the fact that the unit of $|C_n|$ is the same as the unit of the signal amplitude, while the unit of $|X(f)|$ is different from the signal amplitude, which is the amplitude of the signal unit bandwidth.

So $X(f)$ is exactly the spectral density function. In engineering testing, $X(f)$ is still called the spectrum for convenience. The frequency spectrum of general aperiodic signal has the characteristics of continuity and attenuation.

**Example 4-3**　Draw the spectrum of the rectangular window function as shown in Figure 4-15a.

**Solution**: The mathematical description of the rectangular window function in time domain is defined as follows.

$$x(t) = \begin{cases} A & |t| \leqslant \dfrac{\tau}{2} \\ 0 & |t| > \dfrac{\tau}{2} \end{cases}$$

According to the definition of the Fourier transform, its frequency spectrum is calculated as follows.

$$X(\omega) = \int_{-\infty}^{\infty} x(t)\,e^{-j\omega t}\,dt = \int_{-\frac{\tau}{2}}^{\frac{\tau}{2}} A e^{-j\omega t}\,dt$$

$$= \frac{A}{-j\omega}(e^{-j\omega\frac{\tau}{2}} - e^{j\omega\frac{\tau}{2}})$$

$$= A\tau \frac{\sin\left(\omega\dfrac{\tau}{2}\right)}{\omega\dfrac{\tau}{2}}$$

$$= A\tau\,\mathrm{sinc}\left(\frac{\omega\tau}{2}\right)$$

Define the sinc function as

$$\mathrm{sinc}(x) = \frac{\sin x}{x} \tag{4-39}$$

This function is an oscillation function whose cycle is $2\pi$ and decays with the increase of $x$. When the function is at $x = n\pi(n = \pm1, \pm2, \cdots)$, the amplitude is zero, as shown in Figure 4-15b.

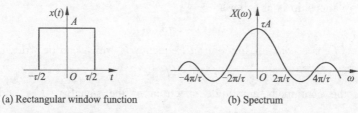

(a) Rectangular window function　　　　(b) Spectrum

**Figure 4-15　Rectangular pulse function and its frequency spectrum**

## 4.3.2.2   The properties of Fourier transform

In signal analysis and processing, Fourier transform is a basic mathematical tool for the transformation between time domain and frequency domain. Mastering the main properties of Fourier transform is helpful to simplify the calculation and analysis of complex signals. The main properties of Fourier transform are listed in Table 4-2, which can be proved by derivation of defined formulas. Here, only a few commonly used properties are described.

**Table 4-2   The main properties of Fourier transform**

| Property name | Time domain | Frequency domain |
| --- | --- | --- |
| Linear superposition | $ax(t)+by(t)$ | $aX(f)+bY(f)$ |
| Symmetry | $\lvert X(f) \lvert x(\pm t)$ | $X(\mp f)$ |
| Scaling | $x(kt)$ | $\dfrac{1}{k}X(\dfrac{f}{k})$ |
| Time-shifting characteristics | $x(t\pm t_0)$ | $X(f)\,\mathrm{e}^{\pm \mathrm{j}2\pi f t_0}$ |
| Frequency shift characteristics | $x(t)\,\mathrm{e}^{\mp \mathrm{j}2\pi f_0 t}$ | $X(f\pm f_0)$ |
| Differential characteristics | $\dfrac{\mathrm{d}^n x(t)}{\mathrm{d}t^n}$ | $(\mathrm{j}2\pi f)^n X(f)$ |
| Integral characteristics | $\displaystyle\int_{-\infty}^{t} x(t)\,\mathrm{d}t$ | $\dfrac{1}{\mathrm{j}2\pi f}X(f)$ |
| Time-domain convolution | $x(t)*y(t)$ | $X(f)Y(f)$ |
| Frequency domain convolution | $x(t)y(t)$ | $X(f)*Y(f)$ |

(1) Linear superposition property

If $X(\omega)=F[x(t)]$, $Y(\omega)=F[y(t)]$ and $a,b$ are constants, then

$$F[ax(t)+by(t)] = aX(\omega)+bY(\omega)$$

This property shows that the Fourier transform is suitable for the analysis of linear systems, and the superposition in the time domain corresponds to the superposition in the frequency domain.

(2) Scale change nature

If $X(f)=F[x(t)]$, and $k$ is a constant greater than zero, then

$$F[x(kt)] = \frac{1}{k}X\left(\frac{f}{k}\right)$$

The time scale characteristic shows that when the signal is expanded in the time domain ($k<1$), the corresponding frequency-domain scale is compressed and the amplitude increases; when the signal is compressed in the time domain ($k>1$), the corresponding frequency-domain scale is expanded and the amplitude decreases. As shown in Figure 4-16.

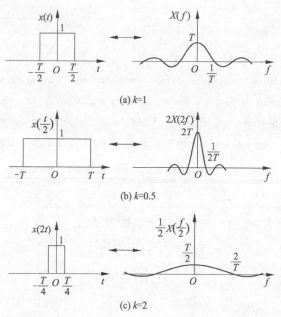

(a) $k=1$

(b) $k=0.5$

(c) $k=2$

**Figure 4-16　Diagram of the nature of scale changes**

(3) Time shifting property

If $t_0$ is constant, then

$$F[x(t\pm t_0)] = X(f)e^{\pm j2\pi f t_0}$$

This property indicates that in the time domain, when the signal shifts along the time axis by a constant value $t_0$, the frequency spectrum function will be multiplied by a factor $e^{\pm j2\pi f t_0}$, that is, only the phase spectrum is changed, and the amplitude spectrum will not be changed, as shown in Figure 4-17.

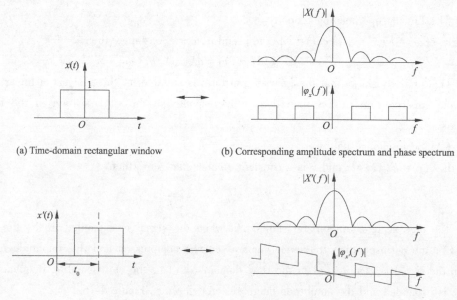

(a) Time-domain rectangular window

(b) Corresponding amplitude spectrum and phase spectrum

(c) Time-domain rectangular window with time shift $t_0$

(d) Corresponding amplitude spectrum and phase spectrum

**Figure 4-17　Examples of time-shifting property**

(4) Convolution property

$$x_1(t) * x_2(t) \Leftrightarrow X_1(f) X_2(f)$$

In the same way

$$x_1(t) x_2(t) \Leftrightarrow X_1(f) * X_2(f)$$

This property shows that: the convolution of two functions in the time domain corresponds to their product in the frequency domain, the product of two functions in the time domain corresponds to their convolution in the frequency domain. Usually, the integral calculation of convolution is difficult, but the use of convolution properties can greatly simplify signal analysis. Therefore, the convolution properties (also known as the convolution theorem) play an important role in signal analysis and classical control theory.

### 4.3.3　Frequency spectrum of typical signal

4.3.3.1　Frequency spectrum of unit impulse function ($\delta$ function)

(1) Definition of $\delta$ function

The rectangular pulse $G(t)$ shown in Figure 4-18 has the width of $\tau$, the height of $1/\tau$, and the area is 1. Keeping the area of the pulse constant, and gradually decrease $x(t)$, then the pulse amplitude will gradually increase. When $\tau \to 0$, the limit of the rectangular pulse is called $\delta$ function, denoted as $\delta(t)$. The $\delta$ function is also called the unit impulse function. The characteristics of $\delta(t)$ are:

From the viewpoint of the limit value of the function, as

$$\delta(t) = \begin{cases} \infty & (t=0) \\ 0 & (t \neq 0) \end{cases} \tag{4-40}$$

From the viewpoint of the area (also commonly referred to as the strength of the $\delta$ function).

$$\int_{-\infty}^{\infty} \delta(t)\,\mathrm{d}t = \lim_{\tau \to 0} \int_{-\infty}^{\infty} G(t)\,\mathrm{d}t = 1 \tag{4-41}$$

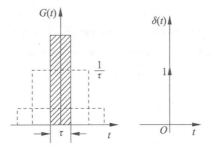

**Figure 4-18　Rectangular pulse and $\delta$ function**

The $\delta$ function is an abstraction of some shocking physical phenomena with extremely short appearance and high energy, such as short-term impact interference in power grid lines, sampling pulses in digital circuits, momentary forces of mechanics, and sudden materials fracture and impact, explosion, etc., these phenomena are analyzed by the $\delta$ function in signal processing. It's just that the function area (energy or intensity) is not necessarily 1, but a constant. Due to the introduction of the $\delta$ function and the use of generalized function theory, the Fourier transform can be extended to the power finite signal that does not satisfy the absolute integrability condition.

(2) The nature of the $\delta$ function

a. Product (sampling) characteristic

If the function $x(t)$ is continuous at $t = t_0$, then

$$x(t)\delta(t) = x(0)\delta(t) \tag{4-42}$$

$$x(t)\delta(t \pm t_0) = x(\mp t_0)\delta(t \pm t_0) \tag{4-43}$$

b. Screening characteristic

Suppose $x(t)$ is a continuous and bounded signal at $t = 0$. When the unit impulse function $\delta(t)$ is multiplied by the signal $x(t)$, the integral of the product $x(0)$ is obtained only at $T_{y1}$, and the product and integral of the other points are zero, thus

$$\int_{-\infty}^{+\infty} x(t)\delta(t)\,\mathrm{d}t = \int_{-\infty}^{+\infty} x(0)\delta(t)\,\mathrm{d}t = x(0)\int_{-\infty}^{+\infty}\delta(t)\,\mathrm{d}t = x(0) \tag{4-44}$$

Similarly, for

$$\int_{-\infty}^{+\infty} x(t)\,\delta(t - t_0)\,\mathrm{d}t = x(t_0)\int_{-\infty}^{+\infty}\delta\,(t - t_0)\,\mathrm{d}t = x(t_0) \tag{4-45}$$

Equations (4-44) and (4-45) show that when the continuous function $x(t)$ is multiplied by the unit pulse signal $\delta(t)$ or $\delta(t - t_0)$, and integrated in the time $(-\infty, \infty)$, the function value of $x(t)$ at $t = 0$ or the function value $x(t_0)$ at $t = t_0$ can be obtained, that is, $x(0)$ or $x(t_0)$ can be filtered out.

c. Convolution characteristic

The convolution between any continuous signal $x(t)$ and $\delta(t)$ is the simplest convolution, and the result is the continuous signal $x(t)$, namely

$$x(t) * \delta(t) = \int_{-\infty}^{+\infty} x(\tau)\delta(t - \tau)\,\mathrm{d}\tau = x(t) \tag{4-46}$$

Similarly, for the time delay unit pulse $\delta(t \pm t_0)$, there is

$$x(t) * \delta(t \pm t_0) = \int_{-\infty}^{+\infty} x(\tau)\delta(t \pm t_0 - \tau)\,\mathrm{d}\tau = x(t \pm t_0) \tag{4-47}$$

The graph of the convolution between continuous signal and the $\delta(t \pm t_0)$ function is shown in Figure 4-19. It can be seen that the geometric meaning of the convolution between the signal $x(t)$ and $\delta(t \pm t_0)$ function is to delay the signal $x(t)$ by $\pm t_0$.

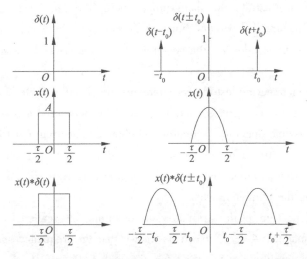

**Figure 4-19 Convolution of continuous signal and δ function**

d. Frequency spectrum of δ function

Take the Fourier transform of the δ function

$$\delta(f) = \int_{-\infty}^{+\infty} \delta(t) e^{-j2\pi ft} dt = e^{-j2\pi f \cdot 0} = 1 \tag{4-48}$$

Its inverse transformation is

$$\delta(t) = \int_{-\infty}^{+\infty} 1 \cdot e^{j2\pi ft} df \tag{4-49}$$

It can be seen from the above equation that the frequency spectrum of the δ function is constant, indicating that the signal contains all the frequency components of $(-\infty, +\infty)$, and the spectral density function of any frequency is equal, as shown in Figure 4-20. This spectrum is often called "uniform spectrum" or "white noise".

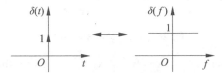

**Figure 4-20 Function δ and frequency spectrum**

The δ function is an even function, that is $\delta(t) = \delta(-t)$ and $\delta(f) = \delta(-f)$. Using the properties of the Fourier transform, such as symmetry, time shift, and frequency shift, etc, a commonly used Fourier transform pair can be obtained.

$$\delta(t \pm t_0) \Leftrightarrow e^{\pm j2\pi f t_0} \tag{4-50}$$

$$e^{\pm j2\pi f_0 t} \Leftrightarrow \delta(f \mp f_0) \tag{4-51}$$

### 4.3.3.2 Frequency spectrum of rectangular window function and constant value function

(1) Frequency spectrum of the rectangular window function

In Example 4-3, the frequency spectrum of the rectangular window function has been calculated and used to illustrate the main properties of the Fourier transform. It should be emphasized that the rectangular window function is evaluated in a finite interval in the time domain, but in the frequency domain the spectrum is continuous and extends indefinitely along the frequency axis.

Since the actual engineering test always intercepts the signal of finite length (window width) in the time domain, its essence is the multiplication of the measured signal and the rectangular window function in the time domain. Therefore, the obtained spectrum must be the convolution of the measured signal spectrum and the rectangular window function spectrum in the frequency domain, so the spectrum obtained by the actual engineering test will also be continuous and infinitely extending on the frequency axis.

(2) Frequency spectrum of the constant value function (also called DC component)

According to Equation (4-48), it can be known that the frequency spectrum of a constant value function with an amplitude of 1 is the $\delta$ function with $f=0$. In fact, the same conclusion can be drawn by using the properties of the Fourier transform. When the window width of the rectangular function is $T \to \infty$, the rectangular function becomes a constant value and its corresponding frequency function tends to be $\delta$ function.

### 4.3.3.3  Unit step signal and its spectrum

As shown in Figure 4-21, the unit step signal $u(t)$ can be expressed as

$$u(t) = \begin{cases} 1 & (t \geqslant 0) \\ 0 & (t < 0) \end{cases} \quad\quad (4\text{-}52)$$

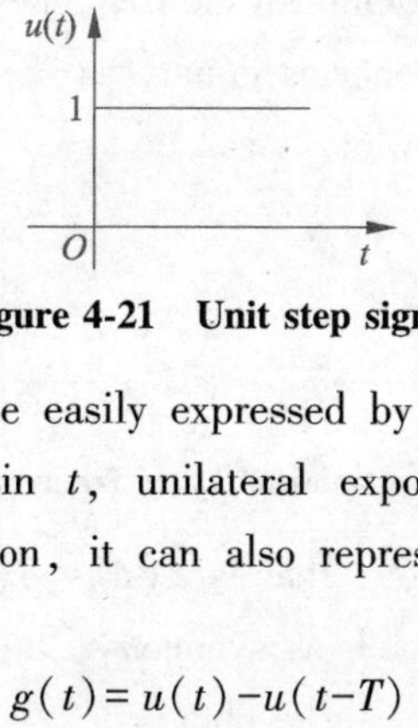

**Figure 4-21  Unit step signal**

Various unilateral signals can be easily expressed by using the unit step signal, such as unilateral sinusoidal signal $u(t) \sin t$, unilateral exponential attenuated oscillation signal $u(t) A e^{-|\alpha| t} \sin 2\pi\omega_0 t$, etc. In addition, it can also represent a single-sided rectangular pulse signal.

$$g(t) = u(t) - u(t-T)$$

Where $T$ is the duration of the rectangular pulse.

Since the unit-step signal does not satisfy the absolute integrability condition, it is not possible to directly obtain the frequency spectrum by Fourier transformation. It can be regarded as the limit of the exponential signal $e^{-\alpha t}$ with $\alpha \to 0$ in the time domain, and its spectrum is the limit of the spectrum of $e^{-\alpha t}$ with $\alpha \to 0$, which is

$$u(t) = \begin{cases} 1 & (t \geqslant 0) \\ 0 & (t < 0) \end{cases} = \begin{cases} \lim\limits_{\alpha \to 0} e^{-\alpha t} & (\alpha > 0, \ t \geqslant 0) \\ 0 & (t < 0) \end{cases}$$

Then

$$x(t) = \begin{cases} \lim_{\alpha \to 0} e^{-\alpha t} & (\alpha > 0, \ t \geqslant 0) \\ 0 & (t < 0) \end{cases}$$

$$X(f) = \lim_{\alpha \to 0} \int_0^{\infty} e^{-\alpha t} \cdot e^{-j2\pi ft} dt$$

$$= \lim_{\alpha \to 0} \frac{1}{\alpha + j2\pi f}$$

$$= \lim_{\alpha \to 0} \frac{\alpha}{\alpha^2 + (2\pi f)^2} - j \frac{2\pi f}{\alpha^2 + (2\pi f)^2}$$

When $f \neq 0$, there is

$$X(f) = -j \frac{1}{2\pi f}$$

When $f = 0$, there is

$$X(f) = \lim_{\alpha \to 0} \frac{\alpha}{\alpha^2 + (2\pi f)^2} - j \frac{2\pi f}{\alpha^2 + (2\pi f)^2}$$

$$= \lim_{\alpha \to 0} \frac{1}{\alpha} \longrightarrow \infty$$

It shows that there is an impact at the frequency $f = 0$, and its intensity is

$$\lim_{\alpha \to 0} \int_{-\infty}^{\infty} X(f) df = \lim_{\alpha \to 0} \int_{-\infty}^{\infty} \frac{\alpha}{\alpha^2 + (2\pi f)^2} df$$

$$= \frac{1}{2\pi} \lim_{\alpha \to 0} \int_{-\infty}^{\infty} \frac{1}{1 + \left(\frac{2\pi f}{\alpha}\right)^2} d \frac{2\pi f}{\alpha}$$

$$= \frac{1}{2\pi} \lim_{\alpha \to 0} \left[ \arctan \frac{2\pi f}{\alpha} \right]_{-\infty}^{\infty}$$

$$= \frac{1}{2}$$

According to the definition and frequency spectrum of the $\delta$ function, it can be seen that the frequency spectrum of the function is

$$X(f) = \frac{1}{2} \delta(f)$$

Therefore, the frequency spectrum of the unit step signal is

$$X(f) = \frac{1}{2} \delta(f) - j \frac{1}{2\pi f} \tag{4-53}$$

The frequency spectrum is shown in Figure 4-22. Since the step signal contains a DC component, the frequency spectrum of the step signal has a pulse at $f = 0$, and it abruptly changes at $t = 0$, so there are high frequency components in the frequency spectrum.

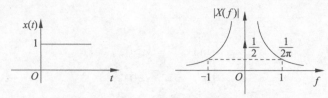

**Figure 4-22　Unit step signal and its spectrum**

### 4.3.3.4　Frequency spectrum of the harmonic function

(1) Frequency spectrum of the cosine function

Using Euler's equation, the cosine function can be expressed as follows

$$x(t) = \cos 2\pi f_0 t = \frac{1}{2}(e^{-j2\pi f_0 t} + e^{j2\pi f_0 t}) \tag{4-54}$$

Its Fourier transform is

$$X(f) = \frac{1}{2}[\delta(f+f_0) + \delta(f-f_0)] \tag{4-55}$$

(2) Frequency spectrum of the sine function

Also using Euler's equation and its Fourier transform is as follows

$$x(t) = \sin 2\pi f_0 t = \frac{j}{2}(e^{-j2\pi f_0 t} - e^{j2\pi f_0 t}) \tag{4-56}$$

$$X(f) = \frac{j}{2}[\delta(f+f_0) - \delta(f-f_0)] \tag{4-57}$$

According to the even-odd and virtual-real properties of the Fourier transform, the cosine function is a real-even function both in time domain and frequency domain; the sine function is a real-odd function in the time domain and an virtual-odd function in the frequency domain. As shown in Figure 4-23.

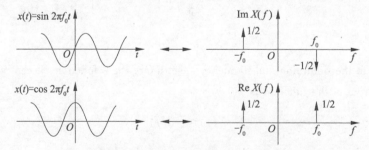

**Figure 4-23　Harmonic function and its spectrum**

### 4.3.3.5　Frequency spectrum of the periodic unit pulse sequence

The pulse sequence of the equally spaced periodic unit is often called a comb function, denoted by $f_{comb}(t, T_s)$, that is

$$f_{comb}(t, T_s) = \sum_{n=-\infty}^{\infty} \delta(t - nT_s) \tag{4-58}$$

Where $T_s$ is the cycle; $n$ is an integer number, $n=\pm1,\pm2,\cdots$.

Because this function is a periodic function, it can be expressed as a complex exponential expansion of Fourier series

$$f_{\text{comb}}(t,T_s) = \sum_{k=-\infty}^{\infty} c_k e^{\text{j}2\pi k f_s t} \tag{4-59}$$

Where $f_s = 1/T_s$ and the coefficient $c_k$ are

$$c_k = \frac{1}{T_s} \int_{-\frac{T_s}{2}}^{\frac{T_s}{2}} f_{\text{comb}}(t,T_s) e^{-\text{j}2\pi f_s t} \text{d}t$$

Because there is only one $\delta$ function in the interval of $\left(-\dfrac{T_s}{2},\dfrac{T_s}{2}\right)$, and when $t=0$, $e^{-\text{j}2\pi f_s t} = e^0 = 1$, so

$$c_k = \frac{1}{T_s} \int_{-\frac{T_s}{2}}^{\frac{T_s}{2}} \delta(t) e^{-\text{j}2\pi f_s t} \text{d}t = \frac{1}{T_s}$$

$$f_{\text{comb}}(t,T_s) = \frac{1}{T_s} \sum_{k=-\infty}^{\infty} e^{\text{j}2\pi k f_s t}$$

And according to Equation (4-51), there is

$$e^{\text{j}2\pi k f_s t} \Leftrightarrow \delta(f-kf_s)$$

The spectrum $f_{\text{comb}}(f,f_s)$ of $f_{\text{comb}}(t,T_s)$ can be obtained, as shown in Figure 4-24, which is also a comb function.

$$f_{\text{comb}}(f,f_s) = \frac{1}{T_s} \sum_{k=-\infty}^{\infty} \delta(f-kf_s) = \frac{1}{T_s} \sum_{k=-\infty}^{\infty} \delta\left(f-\frac{k}{T_s}\right) \tag{4-60}$$

As can be seen from Figure 4-24, the spectrum of the periodic unit pulse sequence is also a periodic pulse sequence.

**Figure 4-24   Periodic unit pulse sequence and its frequency spectrum**

# 4.4   Description of random signal

The random signal is a kind of signal often appears in mechanical engineering. Its characteristics are as follows.

(1) The function cannot be described by precise mathematical equation;

(2) It cannot predict its accurate value at any time in the future;

（3）The results of each observation of this signal are different. A large number of repeated experiments show that it has statistical regularity, so it can be described and studied by probability and statistics methods.

In engineering application, random signals can be seen everywhere, such as the change of temperature, the change of machine vibration, etc. Even if the same machine tool and the same worker processes the same parts, their sizes are not the same. Figure 4-25 shows the strain change of a point on the main beam when the car is driving on a level asphalt road. It can be seen that the sample records of each time point are completely different under the same working conditions (speed, road surface, driving conditions, etc.), and this kind of signal is a random signal.

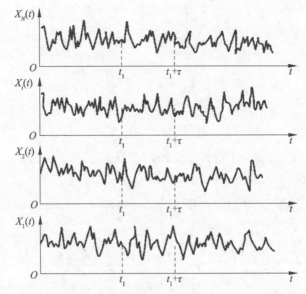

**Figure 4-25   Sample function of the random process**

The physical phenomenon that generates random signals is called random phenomenon. A single $x_i(t)$ representing a random signal is called a sample function. The set $\{x(t)\} = \{x_1(t), x_2(t), \ldots, x_i(t), \ldots, x_N(t)\}$ (also called the whole) of all sample functions that may be generated by a random phenomenon is called a random process.

The statistical characteristics of the random process at any time $t_k$ need to be described by the means of ensemble average. The so-called ensemble average is to add up the value $x_i(t)$ of all samples at a certain time and then devide by the total number of sample functions. For example, in order to find the mean value at $t_1$ in Figure 4-25 we need to add up the value $\{x(t_1)\}$ of all samples at $t_1$ and devide by the number of samples $N$, that is

$$\mu_x(t_1) = \lim_{N \to \infty} \frac{1}{N} \sum_{k=1}^{N} x_k(t_1) \tag{4-61}$$

The correlation of the random process at two time points of $t_1$ and $t_1 + \tau$ can be expressed by the correlation function as follows.

$$R_x(t_1, t_1 + \tau) = \lim_{N \to \infty} \frac{1}{N} \sum_{k=1}^{N} x_k(t_1) x_k(t_1 + \tau) \tag{4-62}$$

In general, both $\mu_x(t_1)$ and $R_x(t_1,t_1+\tau)$ change with time, and this kind of random process is non-stationary random process. If the statistical characteristic parameters of a random process do not change with time, it is called a stationary random process. If the time average statistical characteristics of any sample which belong to the stationary random process are the same and equal to the overall statistical characteristics, then the process is called the ergodic process. As shown in Figure 4-25, the time average of the $i$-th sample is

$$\mu_x(i) = \lim_{T \to \infty} \frac{1}{T} \int_0^T x_i(t)\,\mathrm{d}t = \mu_x \tag{4-63}$$

$$R_x(\tau,i) = \lim_{T \to \infty} \frac{1}{T} \int_0^T x_i(t)x_i(t+\tau)\,\mathrm{d}t = R_x(\tau) \tag{4-64}$$

Most random signals appeared in engineering have ergodicity. Although some are not strictly ergodic processes, they can also be treated as ergodic random processes. Theoretically, it is difficult to obtain the statistical parameters of stochastic process with infinite number of samples. In practice, the random signal is usually treated as each state ergodic process and the aggregate mean value of the whole process is estimated by the time mean value of the finite functions measured. Strictly speaking, only the stationary stochastic process can be ergodic, and only when it is proved that the stochastic process is ergodic, can the set average statistic of the stochastic process be replaced by the sample average statistic.

The main statistical parameters that are usually used to describe the ergodic random signal are: mean, mean square, variance, probability density function, correlation function, etc. The mean value, mean square value, and variance are shown in Equations (4-11), (4-13) and (4-16) respectively. The following only introduces the probability density function, and the signal processing using correlation function will be described in Chapter 6.

### 4.4.1　Probability density function

Figure 4-26 shows the random signal $x(t)$. The probability that the amplitude falls within the range of $x$ and $x+\Delta x$ can be expressed by Equation (4-65).

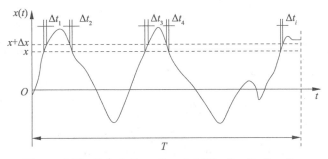

Figure 4-26　Calculation of probability density function

$$P[x \leqslant x(t) < x+\Delta x] = \lim_{T \to \infty} \frac{\Delta t}{T} \tag{4-65}$$

In other words, the probability of $x(t)$ falling within $(x, x+\Delta x)$ can be determined by the limit of the $\Delta t/T$. Where, $\Delta t = \sum_{i=1}^{n} \Delta t_i$ is the total time that $x(t)$ falls in the interval of $(x,x+\Delta x)$; $T$ is the total observation time.

Referring to Figure 4-27a, for the ergodic random signal, the probability that the value of $x(t)$ is less than or equal to the amplitude $\xi$ is

$$P(x)=P[x(t)\leqslant\xi]=\lim_{T\to\infty}\frac{\Delta t[x(t)\leqslant\xi]}{T} \tag{4-66}$$

It is called the probability distribution function.

Since $\xi$ must have a certain lower limit (it can be negative infinity) that $x(t)$ is always greater than it, therefore, as $\xi$ becomes smaller and smaller, the value of the probability distribution function $P(x)$ will always reach zero. Similarly, since the value of $\xi$ must have an upper limit that $x(t)$ can never exceed it, the value of $P(x)$ will always reach 1 when $\xi$ becomes larger and larger. Therefore, the probability distribution function $P(x)$ changes from 0 to 1, as shown in Figure 4-27b.

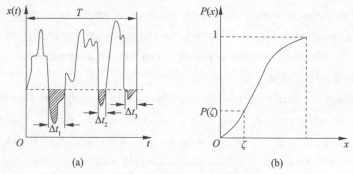

Figure 4-27　Probability distribution function

Although the variation of the probability distribution function is limited to vary in the range of $[0, 1]$, there can be different shapes representing signals with different probability structures. In order to distinguish, generally the slope of the distribution function is used to describe the difference of its probability structure.

$$p(x)=\frac{\mathrm{d}P(x)}{\mathrm{d}x} \tag{4-67}$$

The function obtained in this way is called the probability density function. The change curve is shown in Figure 4-27, and Equation (4-67) can also be written as

$$p(x)=\lim_{\Delta x\to0}\frac{P(x+\Delta x)-P(x)}{\Delta x} \tag{4-68}$$

Where $P(x)$ is the probability distribution function with the instantaneous value of $x(t)$ less than the level of $x$; $P(x+\Delta x)$ is the probability distribution function with the instantaneous value of $x(t)$ less than the level of $(x+\Delta x)$.

According to the probability expression in Equation (4-65), we can know that

$$P[x\leqslant x(t)<x+\Delta x]=P(x+\Delta x)-P(x)=\lim_{T\to\infty}\frac{\Delta t}{T}$$

Therefore, the probability density function or probability density curve $p(x)$ can be written as follows

$$p(x) = \lim_{\Delta x \to 0} \frac{1}{\Delta x} \lim_{T \to \infty} \frac{\Delta t}{T} \qquad (4\text{-}69)$$

The probability density function comprehensively describes the distribution of the instantaneous values of the random process. From Figure 4-28, the area $p(x)\,\mathrm{d}x$ under the $p(x)$ curve is the probability that the instantaneous amplitude falls within $(x, x + \Delta x)$. $p(x)$ is not affected by the size of the amplitude interval, that is, the probability density function represents the change rate of the probability relative to the amplitude, or the probability of unit amplitude, so there is the concept of density, the unit is $1/\Delta x$, and the probability distribution function can also be obtained from the curve of $p(x)$, that is

$$P(x) = \int_{-\infty}^{x} p(x)\,\mathrm{d}x \qquad (4\text{-}70)$$

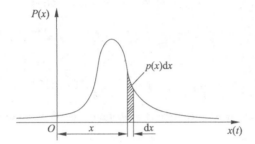

**Figure 4-28　Probability density function curve**

## 4.4.2　Probability density function of typical signal

The probability density functions associated with actual physical phenomena are infinite in number, but as long as the probability density functions of the following three typical signals are mastered, most of the data can be fully approximated. The three probability density functions are the probability density function of normal (Gaussian) noise, the probability density function of the sine wave, the probability density function of the sine wave in noise.

### 4.4.2.1　Normal (Gaussian) noise

The data describing many random physical phenomena in practice can almost be accurately approximated with the following probability density function

$$p(x) = \frac{1}{\sqrt{2\pi} \cdot \sigma_x} \exp\left[ -\frac{(x - \mu_x)^2}{2\sigma_x^2} \right] \qquad (4\text{-}71)$$

Where $\mu_x$ and $\sigma_x$ are the mean value and standard deviation of the data respectively.

The above formula is called the normal or Gaussian probability density function. Gaussian probability density curve and probability distribution curve are shown in Figure 4-29. Its characteristics are as follows:

(1) Single peak, at the peak $x = \mu_x$, the curve takes the x-axis as the asymptote. When $x \to$

$\pm\infty$ , $p(x)\to 0$.

(2) The curve takes $x=\mu_x$ as the axis of symmetry.

(3) $x=\mu_x\pm\sigma_x$ is the inflexion point of the curve.

(4) $P[\mu_x-\sigma_x\leqslant x(t)<\mu_x+\sigma_x]\approx 0.68$

$P[\mu_x-2\sigma_x\leqslant x(t)<\mu_x+2\sigma_x]\approx 0.95$

$P[\mu_x-3\sigma_x\leqslant x(t)<\mu_x+3\sigma_x]\approx 0.995$

The central limit theorem is particularly important in the signal analysis of normal distribution. This theorem can be stated as : if a random variable $x(t)$ is the linear sum of $N$ statistically independent random variables $U_i$, regardless of the probability density of these variables, the probability density of $x=x_1+x_2+\cdots+x_N$ will tend to be normal form when $N$ tends to infinity. Since most physical phenomena are the sum of many random events, the normal formula can provide a reasonable approximation to the probability density function of random data.

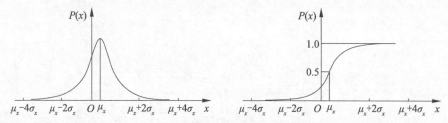

**Figure 4-29　Probability density curve and probability distribution curve of Gaussian signal**

## 4.4.2.2　Sinusoidal signal

For a sinusoidal signal, since the precise amplitude of any future instant can be completely determined by $x(t)=A\sin(2\pi ft+\varphi)$, it is theoretically unnecessary to study its probability distribution. However, if the phase angle $\varphi$ is assumed to be a random variable that obeys a uniform distribution among $\pm\pi$, the sine function can be regarded as a random process. Assuming that its mean value is zero, it can be proved that the probability density function of a sine random process is as follows.

$$p(x)=\begin{cases}\dfrac{1}{\pi}\sqrt{(2\sigma_x^2-x^2)^{-1}} & |x|<A \\ 0 & |x|\geqslant A\end{cases}\qquad(4\text{-}72)$$

Where $\sigma_x=\dfrac{A}{\sqrt{2}}$ is the standard deviation of the sinusoidal signal.

When $\sigma_x=1$, the normalized probability density function of the sinusoidal signal is shown in Figure 4-30. Based on the foregoing knowledge, the probability density can be regarded as the result of the probability limit calculation where $x(t)$ falls within $\Delta x$, that is, the proportion of the time that $x(t)$ falls within $\Delta x$. It can be seen from Figure 4-30 that, for any given $\Delta x$, the sinusoidal signal in each cycle occupies the most time at the peak $\pm A$, while the least time at the mean $\mu_x=0$.

Similar to Gaussian random noise, the probability density function of a sinusoidal signal is completely determined by the mean value and standard deviation. But unlike Gaussian noise, the probability density of a sinusoidal signal is the minimum at the mean value, while the Gaussian noise is the largest.

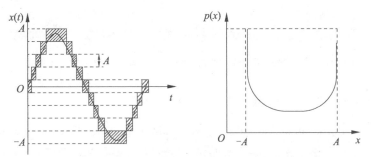

**Figure 4-30 Probability density function of a sine wave**

**Example 4-4** Given the sinusoidal signal $x(t) = A\sin(\omega_0 t + \varphi)$, try to find its probability density function $p(x)$, probability distribution function $P(x)$, mean value $\mu_x$, mean square value $\psi_x$, and variance $\sigma_x^2$.

**Solution:**

$$x(t) = A\sin(\omega_0 t + \varphi)$$

$$(\omega_0 t + \varphi) = \arcsin\frac{x}{A}$$

$$\frac{\mathrm{d}t}{\mathrm{d}x} = \frac{1}{\omega_0}\frac{1/A}{\sqrt{1-(x/A)^2}} = \frac{1}{\omega_0\sqrt{A^2-x^2}}$$

Within a period of one cycle $\left(T = \dfrac{2\pi}{\omega_0}\right)$, as shown in Figure 4-31.

$$p(x) = \lim_{\Delta x \to 0}\frac{1}{\Delta x}\lim_{T \to \infty}\frac{\Delta t}{T} = \frac{1}{T}\cdot\frac{2\mathrm{d}t}{\mathrm{d}x} = \frac{1}{T}\cdot\frac{2\mathrm{d}x}{\omega_0 T\sqrt{A^2-x^2}} = \frac{1}{\pi\sqrt{A^2-x^2}}$$

$$P(x) = \int_{-\infty}^{\infty} p(x)\,\mathrm{d}x = \int_{-A}^{A}\frac{1}{\pi\sqrt{A^2-x^2}} = \left[\arcsin\frac{x}{A}\right]_{-A}^{A} = 1$$

$$\mu_x = \int_{-\infty}^{\infty} xp(x)\,\mathrm{d}x = \int_{-A}^{A}\frac{1}{\pi\sqrt{A^2-x^2}} = \left[-\frac{\sqrt{A^2-x^2}}{\pi}\right]_{-A}^{A} = 0$$

$$\psi_x^2 = \int_{-\infty}^{\infty} x^2 p(x)\,\mathrm{d}x = \int_{-A}^{A}\frac{x^2}{\pi\sqrt{A^2-x^2}} = \frac{1}{\pi}\left[x\sqrt{A^2-x^2} + \frac{x}{2}\sqrt{A^2-x^2} + \frac{A^2}{2}\arcsin\frac{x}{A}\right]_{-A}^{A}$$

$$= \frac{1}{\pi}\cdot\frac{\pi A^2}{2} = \frac{A^2}{2}$$

$$\sigma_x^2 = \psi_x^2 - \mu_x^2 = \frac{A^2}{2}$$

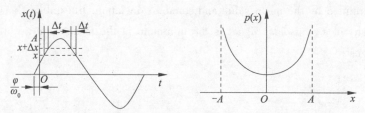

**Figure 4-31   Sine signal and probability density function calculation**

### 4.4.2.3   Sinusoidal signal mixed with Gaussian noise

The expression of a random signal $x(t)$ containing sinusoidal signal $s(t) = S\sin(2\pi ft + \theta)$ is
$$x(t) = n(t) + s(t)$$
Where $n(t)$ is Gaussian random noise with the mean value being zero, and its standard deviation is $\sigma_n$; the standard deviation of $s(t)$ is $\sigma_s$.

The expression of probability density is

$$p(x) = \frac{1}{\sigma_n \pi \sqrt{2\pi}} \int_0^\pi \exp\left[ -\left( \frac{x - S\cos\theta}{4\sigma_n} \right) \right]^2 d\theta \qquad (4\text{-}73)$$

Figure 4-32 is a graph of the probability density function containing a sine wave random signal, $R = (\sigma_s / \sigma_n)^2$. For different values of $R$, $p(x)$ has different graphs. For pure Gaussian noise, $R = 0$; for a sine wave, $R = \infty$; for Gaussian noise containing sine wave, $0 < R < \infty$. This figure provides a graphical basis for identifying whether there is a sinusoidal signal in a random signal, and how much of each account is in the statistical amplitude.

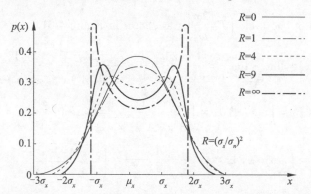

**Figure 4-32   Probability density function of a sinusoidal signal mixed with Gaussian noise**

The probability density function and distribution function of a typical signal are shown in Figure 4-33.

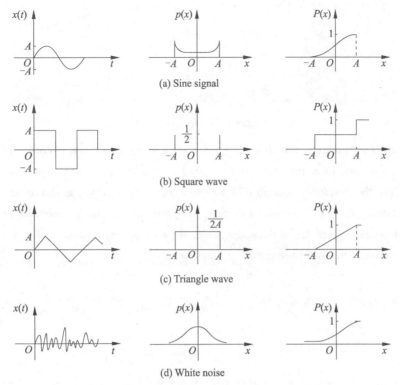

Figure 4-33　**Probability density function and probability distribution function of a typical signal**

# Questions

4.1　Briefly describe several description methods of signals.

4.2　Briefly describe the physical meaning of signal statistical characteristic parameters and their application in mechanical fault diagnosis.

4.3　Write out the mathematical expressions of the two expansions of periodic signals and explain the physical meaning of the coefficients.

4.4　What are the characteristics of the spectrograms of periodic signals and aperiodic signals? What are the similarities and differences in their physical meanings?

4.5　Find the mean value $\mu_x$ and root mean square value $x_{\text{rms}}$ of the sinusoidal signal $x(t) = x_0 \sin \omega t$.

4.6　Using the trigonometric function expansion and complex exponential expansion of the Fourier series, find the frequency spectrum of the periodic triangular wave shown in Figure 4-34 and make a spectrogram.

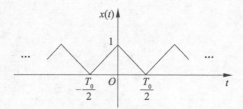

**Figure 4-34    Question 4.6**

4.7    Find the spectrum function $X(f)$, $(a>0, t \geqslant 0)$ of the exponential decay function $x(t) = e^{-at} \cos \omega_0 t$, and draw the graph of the signal in time domain and its spectrum graph.

4.8    Given the frequency spectrum of a certain signal $f(t)$ which is shown in Figure 4-35, find the function $x(t) = f(t) \cos \omega_0 t$ ($\omega_0 > \omega_m$, $\omega_m$ are the angular frequencies of the highest frequency component), and try to draw the graph of its frequency spectrum of $x(t)$. When $\omega_0 < \omega_m$, what happens to the spectrum graph of function $x(t)$?

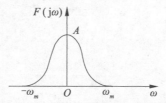

**Figure 4-35    Question 4.8**

4.9    Find the frequency spectrum of the truncated cosine function (shown in Figure 4-36) and make a spectrum graph.

$$x(t) = \begin{cases} \cos \omega_0 t & (\,|t|<T) \\ 0 & (\,|t| \geqslant T) \end{cases}$$

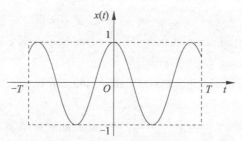

**Figure 4-36    Question 4.9**

4.10    Briefly describe the characteristics of probability density function and its distribution function of typical signal.

# Chapter 5

## Signal Conditioning Method

Although most sensors have converted various measured mechanical quantities into electrical quantities, the output signal of the sensor cannot be directly used for display, transmission, processing and online control due to signal strength is very weak and signal types are complex. Therefore, before using these signals, the signal amplitude, transmission characteristics, and anti-interference ability of the signals must be adjusted according to specific requirements. This chapter will introduce the working principles of bridge circuit, filtering, modulation and demodulation methods commonly used in signal conditioning.

## 5.1 Bridge circuit

When the sensor converts the measured mechanical quantity into the change of parameters of the circuit or magnetic circuit such as resistance, inductance, capacitance, etc., the change of parameters can be further transformed into the change of the output voltage through the bridge circuit. Bridges circuit can be divided into DC bridges and AC bridges according to their different power properties. DC Bridges can only be used to measure changes in resistance, whereas AC Bridges can be used to measure changes in resistance, inductance, and capacitance.

### 5.1.1 Direct current bridge circuit

The bridge that adopts DC power is called DC bridge, and the bridge arm of DC bridge can only be resistance, as shown in Figure 5-1. A bridge circuit powered by the DC power supply is called the DC bridge, whose bridge arms can only be resistance, as shown in Figure 5-1. The resistance $R_1, R_2, R_3$ and $R_4$ are used as the four bridge arms resistances, the DC power supply $U_i$ is connected at both ends of $a$ and $c$, and the voltage $U_0$ is output at both ends of $b$ and $d$.

#### 5.1.1.1 Balance conditions and connection methods of DC bridge

If the load between the output terminals $b$ and $d$ is infinite, that is, when the input impedance of the connected instrument or amplifier is large, it can be regarded as an open circuit. At this time, the current of the bridge circuit is as follows

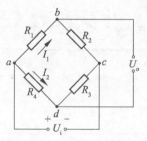

$$I_1 = \frac{U_i}{R_1 + R_2}$$

$$I_2 = \frac{U_i}{R_3 + R_4}$$

**Figure 5-1　DC bridge**

Therefore, the output voltage of the bridge circuit is

$$U_o = U_{ab} - U_{ad} = I_1 R_1 - I_2 R_4$$

$$= \left( \frac{R_1}{R_1 + R_2} - \frac{R_4}{R_3 + R_4} \right) U_i \tag{5-1}$$

$$= \frac{R_1 R_3 - R_2 R_4}{(R_1 + R_2)(R_3 + R_4)} U_i$$

In Equation (5-1), when

$$R_1 R_3 = R_2 R_4 \tag{5-2}$$

The output voltage of the bridge circuit is zero, Equation (5-2) is called the balance condition of DC bridge circuit.

According to the changes in the resistance value of the bridge arm during the operation of the bridge circuit, it can be divided into three connection modes, such as half bridge single arm, half bridge double arm and full bridge, as shown in Figure 5-2.

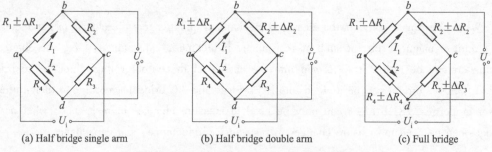

(a) Half bridge single arm　　　(b) Half bridge double arm　　　(c) Full bridge

**Figure 5-2　Connection mode of DC bridge**

Figure 5-2a shows the half-bridge single-arm connection mode. Only one arm's resistance value changes with the change of measured quantity during operation. In the figure, the value of resistance $R_1$ increases by $\Delta R_1$. According to Equation (5-1), the output voltage is

$$U_o = \left( \frac{R_1 + \Delta R_1}{R_1 + \Delta R_1 + R_2} - \frac{R_4}{R_3 + R_4} \right) U_i$$

In practice, in order to simplify the design and obtain the maximum sensitivity of the bridge, the resistance value of two adjacent bridge arms are often equal, that is $R_1 = R_2 = R_0$, $R_3 = R_4 =$

$R_0'$. If $R_0 = R_0'$, the output voltage is

$$U_{\text{o}} = \frac{\Delta R_0}{4R_0 + 2\Delta R_0} U_{\text{i}}$$

Because the change value of the bridge arm resistance is far less than its own resistance value, that is $\Delta R_0 \ll R_0$, so

$$U_{\text{o}} \approx \frac{\Delta R_0}{4R_0} U_{\text{i}} \tag{5-3}$$

The output voltage of the bridge circuit is proportional to the input voltage. Under the condition of $\Delta R_0 \ll R_0$, the output voltage of the bridge circuit is also proportional to $\Delta R_0 / R_0$. The sensitivity of the bridge circuit is defined as

$$S_{\text{B}} = \frac{U_0}{\Delta R_0 / R_0} \tag{5-4}$$

Therefore, the sensitivity of the half-bridge single arm is $S_{\text{B}} \approx \frac{1}{4} U_{\text{i}}$. In order to improve the sensitivity of the bridge circuit, it usually adopts the half bridge double arm in the application, as shown in Figure 5-2b. In this case, there are two bridge arms that change with the measured quantity, that is, $R_1 \rightarrow R_1 \pm \Delta R_1$, $R_2 \rightarrow R_2 \mp \Delta R_2$. When $R_1 = R_2 = R_3 = R_4 = R_0$, $\Delta R_1 = \Delta R_2 = \Delta R_0$, the output of the bridge circuit is

$$U_{\text{o}} = \frac{\Delta R_0}{2R_0} U_{\text{i}} \tag{5-5}$$

Similarly, when the full bridge is adopted in the application, as shown in Figure 5-2c, the resistance values of the four bridge arms change with the measured quantity, that is, $R_2 \rightarrow R_2 \mp \Delta R_2$, $R_3 \rightarrow R_3 \pm \Delta R_3$, $R_4 \rightarrow R_4 \mp \Delta R_4$, $\Delta R_1 = \Delta R_2 = \Delta R_3 = \Delta R_4 = \Delta R_0$, the output of the bridge circuit is

$$U_{\text{o}} = \frac{\Delta R_0}{R_0} U_{\text{i}} \tag{5-6}$$

It can be seen from the above analysis that different bridge circuits have different output voltages. Among them, the full bridge circuit can obtain the maximum output voltage, and its sensitivity is four times that of the half bridge single arm circuit.

## 5.1.1.2　Measurement error and its compensation in the bridge circuit

For the bridge circuit, the error mainly comes from the nonlinear error and the temperature error. It can be seen from Equation (5-3) that when the half bridge single arm connection method is adopted, the output voltage is approximately proportional to $\Delta R_0 / R_0$, which is mainly caused by the nonlinearity of the output voltage. The way to reduce the nonlinear error is to use half bridge double arm or full bridge method, as Equations (5-5) and (5-6) described. At this time, not only the nonlinear error can be eliminated, but the output sensitivity is also doubled.

Another kind of error is the temperature error, which is caused by different resistance changes caused by temperature changes. That is, as in the above two arm electric bridge circuits,

$\Delta R_1 \neq -\Delta R_2$; as in the full bridge method, $\Delta R_1 \neq -\Delta R_2$ or $\Delta R_3 \neq -\Delta R_4$. Therefore, in order to reduce the temperature error when using the resistance strain gauge, the temperature of each strain gauge should be the same as far as possible when pasting the strain gauge, the temperature compensation gauge or half bridge double arm or full bridge should be used.

### 5.1.1.3   The interference of DC bridge circuit

The output voltage of the bridge circuit is the product of $\Delta R_0 / R_0$ and $U_i$. Because $\Delta R_0 / R_0$ is very small, the disturbance caused by the voltage instability in the power supply cannot be ignored. In order to suppress the interference, the following measures are usually adopted:

(1) The signal lead of the bridge is shielded cable;

(2) The shielded metal mesh of the shielded cable should be connected with the negative terminal of the power supply to the bridge circuit, and should be isolated from the enclosure of the amplifier;

(3) The amplifiers should have a high common-mode rejection ratio.

## 5.1.2   Alternating current bridge

It is known from the DC bridge circuit that when the input voltage and resistance are known, the change value of the bridge resistance can be measured through the change of output voltage. When the power supply is an alternating power supply, the equations mentioned in the DC bridge circuits are still true, and the bridge circuit is called AC bridge. When the four bridge arms use capacitors or inductors as elements, the relevant theories of the AC bridge circuit must be applied to analyze the relevant properties of the bridge circuit.

When the capacitance and inductance are written in vector form, Equation (5-2) of the bridge balance condition can be rewritten as

$$\vec{Z}_1 \vec{Z}_3 = \vec{Z}_2 \vec{Z}_4 \qquad (5\text{-}7)$$

When written in complex exponential form, there are

$$\vec{Z}_1 = Z_1 e^{j\varphi_1} \qquad \vec{Z}_2 = Z_2 e^{j\varphi_2}$$

$$\vec{Z}_3 = Z_3 e^{j\varphi_3} \qquad \vec{Z}_4 = Z_4 e^{j\varphi_4}$$

Substituting the above equation into Equation (5-7), there are

$$Z_1 Z_3 e^{j(\varphi_1 + \varphi_3)} = Z_2 Z_4 e^{j(\varphi_2 + \varphi_4)} \qquad (5\text{-}8)$$

The condition for the establishment of this equation is that the impedance modes on both sides of the equation are equal and the impedance angles are equal, that is

$$\begin{cases} Z_1 Z_3 = Z_2 Z_4 \\ \varphi_1 + \varphi_3 = \varphi_2 + \varphi_4 \end{cases} \qquad (5\text{-}9)$$

In Equation (5-9), $Z_1, \cdots, Z_4$ is called the mode of impedance; $\varphi_1, \cdots, \varphi_4$ is called the impedance angle. Therefore, the AC bridge circuit needs to adjust two kinds of balance, one is impedance balance, and the other is impedance angle matching.

There are different combinations of AC bridges, including capacitance and inductance bridges. If the two adjacent arms are connected with resistances, the other two arms are connected with the same impedance, as shown in Figure 5-3.

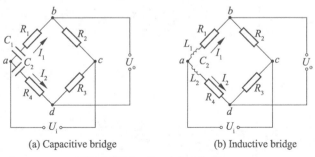

(a) Capacitive bridge                    (b) Inductive bridge

**Figure 5-3   Converter bridge**

For the capacitor bridge shown in Figure 5-3a, it can be seen from Equations (5-7) and (5-8) that the equilibrium condition is

$$\left(R_1+\frac{1}{j\omega C_1}\right) R_3 = \left(R_4+\frac{1}{j\omega C_2}\right) R_2$$

The real and imaginary parts of both sides of the above equation are equal respectively, and the following bridge equilibrium equations can be obtained

$$\begin{cases} R_1 R_3 = R_2 R_4 \\ \dfrac{R_3}{C_1} = \dfrac{R_2}{C_2} \end{cases} \tag{5-10}$$

By comparing Equation (5-2) with Equation (5-10), it can be seen that the first equation in (5-10) is exactly the same as that of Equation (5-2), which means that the balance conditions of capacitive bridge circuit shown in Figure 5-3a must satisfy the balance requirements of both capacitance and resistance.

For the inductive bridge shown in Figure 5-3b, the balance condition is

$$(R_1+j\omega L_1) R_3 = (R_4+j\omega L_2) R_2$$

which is

$$\begin{cases} R_1 R_3 = R_2 R_4 \\ L_1 R_3 = L_2 R_2 \end{cases} \tag{5-11}$$

Figure 5-4 shows the circuit diagram of AC bridge for strain gauge. The unbalance of resistance can be adjusted by selecting the resistance $R_1$ and $R_2$ and variable resistance $R_3$ through switch $S$, and the unbalance of distributed capacitance of bridge arm to ground can be adjusted by differential variable capacitance $C_2$.

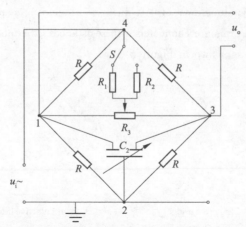

**Figure 5-4　AC resistance bridge with resistors and capacitors**

From the balance condition Equations (5-7) to (5-9) of the AC bridge, and the analysis of the balance conditions of the capacitor bridge and the inductance bridge, it can be seen that these balance conditions are only derived for the case that the power supply of the bridge contains only one frequency $\omega$.

When the bridge power supply has multiple frequency components, the balance condition cannot be obtained, that is, the bridge is unbalanced. Therefore, it is required for the bridge power supply to have good voltage waveform and frequency stability in the AC bridge.

When AC bridge is adopted, we should also pay attention to some parameters that affect the measurement error, such as mutual inductance between components in the bridge, residual reactance of non-inductive resistance, inductive effect of adjacent AC circuit on the bridge, leakage resistance and distributed capacitance between components and ground, etc.

## 5.2　Signal filtering

In the analysis and processing of the measured signal, there is the problem that the useful signal is superimposed with unnecessary noise. Some of these noises are generated simultaneously with the signal, and some are mixed in during signal transmission. The noise is sometimes larger than the useful signal to overwhelm the useful signal. Therefore, removing or reducing the interference noise from the original signal becomes an important issue in signal processing.

According to the different characteristics of the useful signal, the process of eliminating or reducing the interference noise and extracting the useful signal is called filtering, and the system that realizes the filtering function is called a filter. The classic filter is a circuit with frequency selection characteristics. When the noise and useful signal are in different frequency bands, the noise will be greatly attenuated or eliminated by the filter, while the useful signal is retained. But when the frequency of the noise and the useful signal are in the same frequency band, the classic filter cannot achieve the above-mentioned functions. The actual needs have stimulated the development of another kind of filter, that is, starting from the concept of statistics, the extracted signal is estimated in the time domain. In the sense of optimal statistical index, the estimated value is optimal to approximate the useful signal, and the noise is also weakened or eliminated in the sense of optimal statistics. These two kinds of filters are widely used in many fields. This section only discusses the former.

According to the range of the passband and stopband of the amplitude frequency characteristics of the filter, it can be divided into low pass, high pass, band pass, band stop and other types. According to the best approximation characteristics, it can be classified into Butterworth filter, Chebyshev filter, Bessel filter and so on. According to the nature of the signal processed by the filter, it can be divided into analog filter and digital filter. Analog filter is used to process analog signal (continuous time signal), and digital filter is used to process discrete time signal.

## 5.2.1  Ideal analog filter

The ideal analog filter is an idealized model that is physically unrealizable, but the analysis for the ideal analog filter can help to further understand the transmission characteristics of the actual filter. On the one hand, the concepts derived from the ideal filter have universal meaning for the actual filter. On the other hand, we can also use some methods to improve the characteristics of the actual filter, so as to achieve the purpose of approximating the ideal filter. The amplitude-frequency characteristic curve of an ideal analog filter is shown in Figure 5-5.

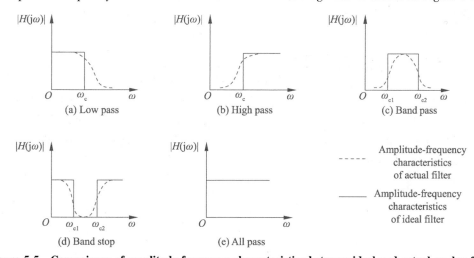

**Figure 5-5  Compairson of amplitude-frequency characteristics between ideal and actual analog filter**

In Figure 5-5, the frequency components of the signal below a certain frequency $\omega_c$ can be passed through with the same magnification by the ideal low-pass filter, and the frequency components above $\omega_c$ will be reduced to zero. $\omega_c$ is called the cut-off frequency of the filter, the frequency range of $\omega < \omega_c$ is called the pass band of the low-pass filter, and the frequency range of $\omega > \omega_c$ is called the stop band. The high pass filter is the opposite to low pass filter, the frequency range of its pass band is $\omega > \omega_c$ and the frequency range of the stop band is $\omega < \omega_c$. The pass band of the band pass filter is between the lower cut-off frequency $\omega_{c1}$ and the upper cut-off frequency $\omega_{c2}$, and the stop band of the band stop filter is between $\omega_{c1}$ and $\omega_{c2}$. For the all-pass filter, the signal of each frequency component can pass through with the same magnification.

The ideal low-pass filter is one of the most common ideal filters with rectangular amplitude frequency characteristics and linear phase characteristics. Because the ideal high pass, band pass and band stop can be obtained by the ideal low-pass series and parallel, we study the time-domain characteristics of the ideal low-pass filter through the unit impulse response.

The ideal low-pass filter has rectangular amplitude-frequency characteristics and linear phase characteristics, which can be expressed as

$$\begin{cases} A(\omega)=1 & |\omega|<\omega_c \\ \varphi(\omega)=-t_0\omega & |\omega|>\omega_c \end{cases} \tag{5-12}$$

The graph is shown in Figure 5-6.

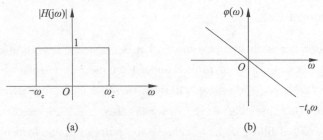

**Figure 5-6　Characteristics of ideal low-pass filter**

Taking the inverse Fourier transform to the frequency response function $H(\omega)=A(\omega)\,e^{j\varphi(\omega)}$, the unit impulse response of the ideal low-pass filter can be obtained as

$$h(t)=\frac{\omega_c}{\pi}\frac{\sin[\omega_c(t-t_0)]}{\omega_c(t-t_0)}=\frac{\omega_c}{\pi}\text{sinc }x[\omega_c(t-t_0)] \tag{5-13}$$

Equation (5-13) shows that the unit impulse response of an ideal low-pass filter is a sampling function sinc $x[\omega_c(t-t_0)]$ delayed by $t_0$, and its waveform is shown in Figure 5-7. Since the impulse response appears before the excitation ($t<0$), the ideal low-pass filter is a non-causal system, which is physically unrealizable.

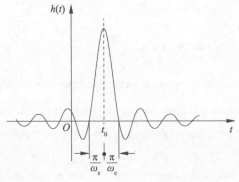

**Figure 5-7　Impulse response of an ideal low-pass filter**

The smaller the cut-off frequency $\omega_c$ of the ideal low-pass, the greater the distortion of its output $h(t)$ compared to the input impulse signal $\delta(t)$. When the cut-off frequency $\omega_c$ of ideal low-pass increases, the first zero point $t\pm\pi/\omega_c$ of impulse response $h(t)$ on both sides of $t=t_0$ is gradually close to $t_0$, and when $\omega_c\to\infty$, then $h(t)\to\delta(t)$. From the perspective of frequency spectrum, the bandwidth of input signal $\delta(t)$ is infinite, but the bandwidth of ideal low-pass filter is limited, so distortion is inevitable.

## 5.2.2 Actual analog filter and its basic parameters

From the previous analysis, it can be seen that the ideal filter is a physically unrealizable system, and the filters used in engineering are not ideal filters. However, actual filters constructed according to certain rules, such as Butterworth filters, Chebyshev filters, and Elliptic filters, have their amplitude-frequency characteristics close to those of an ideal filter. Figure 5-8 shows the amplitude-frequency characteristics of these three types of low-pass filters. Their amplitude-frequency characteristics have the characteristics of flat in passband change, equal in passband fluctuation and equal in stopband and passband fluctuation.

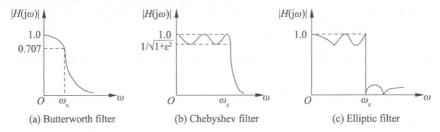

(a) Butterworth filter  (b) Chebyshev filter  (c) Elliptic filter

**Figure 5-8 Amplitude-frequency characteristics of three commonly used low-pass filters**

For an ideal filter, we only need to specify the cut-off frequency to explain its performance, that is to say, the ideal filter can be selected only according to the cut-off frequency, because within the cut-off frequency its amplitude-frequency characteristic is a constant, and outside the cut-off frequency it is zero. For the actual analog filter, its characteristic curve has no obvious turning point, and the passband amplitude is not constant, as shown in Figure 5-9. Therefore, we need more characteristic parameters to describe and select the actual filter. In addition to cut-off frequency, these parameters mainly include ripple amplitude, bandwidth, quality factor and octave selectivity.

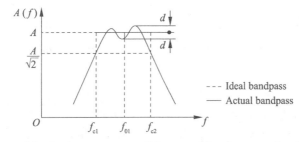

**Figure 5-9 Basic parameters of the actual band-pass analog filter**

(1) Ripple amplitude $d$

Within a certain frequency range, the amplitude-frequency characteristics of the actual filter may show wave-like changes. The smaller the ratio of the fluctuation amplitude $d$ to the average value $A$ of the amplitude frequency characteristics, the better. In general, it should be much less than-3dB, that is to say, there should be $20\lg(d/A) \ll -3$ dB, i.e. $d \ll A/\sqrt{2}$.

(2) Cut-off frequency $f_c$

The frequency corresponding to the amplitude-frequency characteristic value equal to $A/\sqrt{2}$ is called the cut-off frequency of the filter, as shown in Figure 5-9 $f_{c1}$ and $f_{c2}$. If the signal power is expressed by the square of the signal amplitude, then this point is exactly the half power point.

(3) Bandwidth $B$

The frequency range between the upper and lower cut-off frequencies is called filter bandwidth, or $-3$ dB bandwidth, and the unit is Hz. The bandwidth determines the ability of the filter to distinguish the adjacent frequency components in the signal, namely frequency resolution.

(4) Quality factor $Q$

For band-pass filters, the ratio of center frequency $f_c$ to bandwidth $B$ is usually called the quality factor of the filter, that is, $Q = f_c/B$.

The center frequency is defined as the square root of the product of upper and lower cut-off frequencies, that is, $f_c = \sqrt{f_{c1} \cdot f_{c2}}$. The quality factor affects the shape of the amplitude frequency characteristic of the low-pass filter at the cut-off frequency.

(5) Octave selectivity $\lambda$

On the outside of the two cut-off frequencies, the actual filter has a transition zone. The slope degree of the amplitude frequency characteristic curve of the transition zone reflects the attenuation speed of the amplitude frequency characteristic, which determines the ability of the filter to attenuate the frequency components outside the bandwidth.

It is usually characterized by octave selectivity. Octave selectivity is the attenuation value of amplitude frequency characteristics between the upper cut-off frequencies $f_{c2}$ and $2f_{c2}$, or between the lower cut-off frequencies $f_{c1}$ and $f_{c1}/2$, that is, the attenuation value when the frequency changes by one octave, expressed in dB. The faster the attenuation, the better the filter selectivity. The attenuation far away from the cut-off frequency can be expressed by 10 octave attenuation.

Another way to express filter selectivity is to use the ratio of the $-60$ dB bandwidth to the $-3$ dB bandwidth in the filter amplitude-frequency characteristic, that is

$$\lambda = \frac{B_{-60 \text{ dB}}}{B_{-3 \text{ dB}}} \tag{5-14}$$

In the ideal filter, $\lambda = 1$. And in the commonly used filter, $\lambda = 1 \sim 5$. For some filters, due to the influence of its components, the stop band attenuation multiple cannot reach $-60$ dB, so the selectivity is expressed by the ratio of the bandwidth of the marked attenuation multiple (such as $-40$ dB or $-30$ dB) to $-3$ dB bandwidth.

# 5.3    Signal modulation and demodulation

Modulation is a commonly used conditioning method in the transmission of long-distance test signals, mainly to solve the problems of the amplification of weak and slowly changing signals and the long-distance transmission of signals. For example, the measured physical quantities (such as temperature, displacement, force, etc.) are mostly weak signals with low frequency and slow variation. For such a type of signal, it will be difficult to directly send it to the DC amplifier for amplification. This is because the amplification of DC amplifier with direct coupling between stages will be affected by zero drift. When the magnitude of the drift signal is close to or exceeds the measured signal, the measured signal will be submerged by the zero drift after step-by-step amplification.

In order to solve the problem of the amplification of slowly varying signals, a method of signal modulation is adopted in the signal processing, that is, the weak slowly varying signal is first loaded into the high frequency AC signal, then amplified by the AC amplifier, and finally the amplified slowly varying signal is taken out from the amplifier. The process is shown in Figure 5-10. The transformation process in this signal transmission is called modulation and demodulation. In signal analysis, signal truncation and window function weighting are also a kind of amplitude modulation; in sound signal measurement, the superposition, product and convolution of sound signal caused by echo effect, among which the product of sound signal belongs to amplitude modulation.

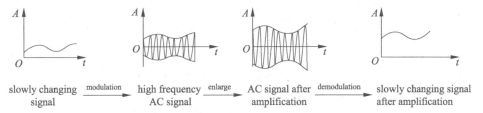

slowly changing signal $\xrightarrow{\text{modulation}}$ high frequency AC signal $\xrightarrow{\text{enlarge}}$ AC signal after amplification $\xrightarrow{\text{demodulation}}$ slowly changing signal after amplification

**Figure 5-10    Signal modulation and demodulation process**

The types of signal modulation can be divided into amplitude modulation(AM), frequency modulation(FM) and phase modulation(PM).

## 5.3.1    Amplitude modulation

### 5.3.1.1    Principles of modulation and demodulation

Amplitude modulation is to multiply a high-frequency sinusoidal signal (or carrier wave) with the test signal, so that the amplitude of the carrier signal changes with the test signal. The cosine signal $z(t)$ with the frequency $f_z$ as carrier is analyzed below.

Based on the properties of Fourier transform, the product of two signals in time domain corresponds to convolution of the two signals in frequency domain. In other words, the Fourier transform of the product of two signals in the time domain, is equal to the convolution of the Fourier transform of each of these signals, that is

$$x(t)z(t) \Leftrightarrow X(f) * Z(f) \tag{5-15}$$

The spectrum of cosine function is a pair of pulse lines, that is

$$z(t) = \cos(2\pi f_z t) \Leftrightarrow \frac{1}{2}\delta(f-f_z) + \frac{1}{2}\delta(f+f_z) \tag{5-16}$$

The result of convolution between a function and a unit impulse function is that the graph is shifted from the origin to the impulse function. Therefore, if the high-frequency cosine signal is used as the carrier, multiply the signal $x(t)$ by the carrier signal $z(t)$, the result is equivalent to shifting the original signal spectrum pattern from the origin to the carrier frequency, and its amplitude is halved, as shown in Figure 5-11. Therefore, the amplitude modulation process is equivalent to the frequency "shifting" process.

$$x_m(t) = x(t) \cdot \cos(2\pi f_z t)$$

$$X_m(f) = \frac{1}{2}X(f) * \delta(f+f_z) + \frac{1}{2}X(f) * \delta(f-f_z) \tag{5-17}$$

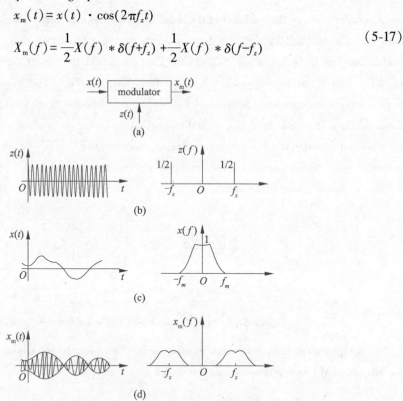

Figure 5-11　Diagram of signal amplitude modulation process

If the amplitude modulated wave $x_m(t)$ is multiplied by the carrier signal $z(t)$ again, the frequency domain figure will also be "shifted" again, that is, the Fourier transform of the product of $x_m(t)$ and $z(t)$ is

$$F[x_m(t)z(t)] = \frac{1}{2}X(f) + \frac{1}{4}X(f) * \delta(f+2f_z) + \frac{1}{4}X(f) * \delta(f-2f_z) \tag{5-18}$$

This result is shown in Figure 5-12. If the high-frequency component with the center frequency of $2f_z$ is filtered out by a low-pass filter, the spectrum of the original signal can be reproduced (only its amplitude is reduced by half, which can be compensated by amplification). This process is called synchronous demodulation. "Synchronization" means that the signal multiplied during demodulation has the same frequency and phase as the carrier signal during modulation.

In the above modulation method, the modulation signal $x(t)$ is directly multiplied by the carrier signal $z(t)$. The amplitude modulation wave has polarity change, that is, when the signal crosses the zero line, its amplitude changes suddenly from positive to negative (or from negative to positive), and its phase (relative to carrier signal) also changes 180° correspondingly. This modulation method is called suppressed amplitude modulation. The amplitude and polarity of the original signal can only be reflected by synchronous demodulation or phase-sensitive demodulation.

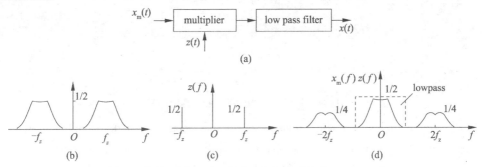

Figure 5-12　Diagram of signal demodulation process

If the modulated signal $z(t)$ is biased and a DC component $A$ is superimposed to make the biased signal have a positive voltage, the expression of AM wave is

$$x_{\mathrm{m}}(t) = [A + x(t)] \cos 2\pi f_z t \tag{5-19}$$

This modulation method is called unsuppressed amplitude modulation, or offset amplitude modulation. The envelope of the amplitude modulation wave has the shape of the original signal, as shown in Figure 5-13a. Generally for unsuppressed AM waves, the original signal can be recovered after rectification and filtering (or called envelope detection method).

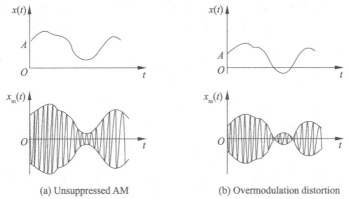

(a) Unsuppressed AM　　　　(b) Overmodulation distortion

Figure 5-13　Unsuppressed AM and overmodulation distortion

### 5.3.1.2 Waveform distortion of AM wave

After the signal is modulated, waveform distortion may occur in the following situations.

(1) Overmodulation distortion

For non-suppressed amplitude modulation, the DC bias must be large enough, otherwise the phase of $x(t)$ will undergo a 180° phase change, as shown in Figure 5-13b, which is called overmodulation. At this time, if the envelope method is used for detection, the detected signal will be distorted, and the original signal cannot be recovered.

(2) Overlapping distortion

AM waves are composed of a pair of double sideband signals with $f_m$ on each side. When the carrier frequency $f_z$ is low, the lower sideband of the positive frequency end will overlap with the upper sideband of the negative frequency end, as shown in Figure 5-14.

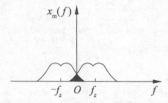

**Figure 5-14    Frequency aliasing effect**

This is similar to the frequency aliasing effect that occurs when the sampling frequency is low. Therefore, it is required that the carrier frequency $f_z$ must be greater than the highest frequency $f_g$ in the modulation signal $x(t)$, that is, $f_z > f_g$. In practical applications, the carrier frequency is often selected to be at least several times or even tens of times the highest frequency in the original signal.

(3) Distortion of AM waves as they pass through the test system

When the AM wave passes through the test system, it will also be distorted by the frequency characteristics of the test system itself.

### 5.3.1.3 Typical AM wave and its spectrum

In order to understand the characteristics of AM waves in the time domain and frequency domain, the frequency spectrum of some typical AM waves are listed in Figure 5-15.

(1) DC modulation

$$x_{\mathrm{m}}(t) = 1 \times \cos(2\pi f_z t)$$

$$X_{\mathrm{m}}(f) = \frac{1}{2} [\delta(f+f_z) + \delta(f-f_z)] \tag{5-20}$$

(2) Cosine modulation

$$x_{\mathrm{m}}(t) = \cos 2\pi f_0 \cos 2\pi f_z t$$

$$X_{\mathrm{m}}(f) = \frac{1}{4} [\delta(f+f_z+f_0) + \delta(f+f_z-f_0) + \delta(f-f_z-f_0) + \delta(f-f_z+f_0)] \tag{5-21}$$

(3) Cosine bias modulation

$$x_m(t) = (\cos 2\pi f_0 + 1)\cos 2\pi f_z t$$

$$X_m(f) = \frac{1}{2}[\delta(f+f_z) + \delta(f-f_z)] + \frac{1}{4}[\delta(f+f_z+f_0) + \delta(f+f_z-f_0)$$

$$+ \delta(f-f_z-f_0) + \delta(f-f_z+f_0)] \tag{5-22}$$

(4) Rectangular pulse modulation

$$x_m(t) = x(t)\cos 2\pi f_z t$$

The expression of rectangular pulse is

$$x(t) = \begin{cases} 1 & |t| \leq \dfrac{\tau}{2} \\ 0 & \text{Others} \end{cases}$$

$$X_m(f) = \frac{\tau}{2}[\text{sinc } \pi(f+f_z)\tau + \text{sinc } \pi(f-f_z)\tau] \tag{5-23}$$

(5) Periodic rectangular pulse modulation

$$x_m(t) = x(t)\cos 2\pi f_z t$$

The expression of periodic rectangular pulse signal in one period and its Fourier transform is

$$x(t) = \begin{cases} 1 & |t| \leq \dfrac{\tau}{2} \\ 0 & |t| > \dfrac{\tau}{2} \end{cases}$$

$$X(f) = 4\pi^2 \sum_{n=-\infty}^{\infty} \frac{A\tau}{T}\text{sinc } \pi n f_0 \tau \delta(f - n f_0)$$

That is, $X_m(f) = X(f) * Z(f)$

$$= 4\pi^2 \sum_{n=-\infty}^{\infty} \frac{A\tau}{T}\text{sinc } \pi n f_0 \tau \delta(f-n f_0) * \frac{1}{2}[\delta(f+f_0) + \delta(f-f_0)]$$

$$= 2\pi^2 \sum_{n=-\infty}^{\infty} \frac{A\tau}{T}[\text{sinc } \pi n f_0 \tau \delta(f-n f_0-f_z) + \text{sinc } \pi n f_0 \tau \delta(f-n f_0+f_z)] \tag{5-24}$$

(6) Bias modulation for arbitrary frequency-limited signal

$$x_m(t) = [1+x(t)]\cos 2\pi f_z t$$

$$X_m(f) = \frac{1}{2}[\delta(f+f_z) + \delta(f-f_z)] + \frac{1}{2}[X(f+f_z) + X(f-f_z)] \tag{5-25}$$

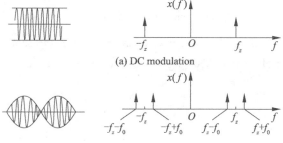

(a) DC modulation

(b) Cosine modulation

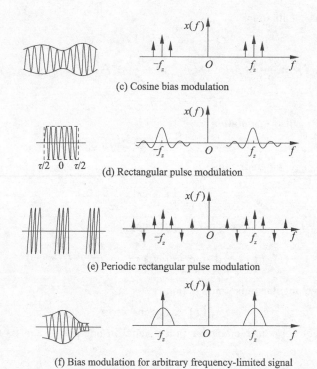

(c) Cosine bias modulation

(d) Rectangular pulse modulation

(e) Periodic rectangular pulse modulation

(f) Bias modulation for arbitrary frequency-limited signal

**Figure 5-15　Waveform and frequency spectrum of a typical AM wave**

### 5.3.1.4　Application of amplitude modulation in test equipment

Figure 5-16 shows the block diagram of the dynamic resistance strain gauge in practical application. In this figure, the resistance strain gauge attached to the test piece produces a corresponding resistance change under the action of external force $x(t)$, and is connected to the bridge. The oscillator generates a high-frequency sinusoidal signal $z(t)$ as the operating voltage of the bridge circuit.

According to the working principle of the bridge circuit, it is equivalent to a multiplier whose output is the product of the signal $x(t)$ and the carrier signal $z(t)$, so the output of the bridge circuit is the amplitude modulation signal $x_m(t)$. After AC amplification, in order to obtain the original waveform of the signal $x(t)$, phase sensitive detection is needed to be applied in the amplitude modulation signal $x_m(t)$, that is, synchronous demodulation.

At this time, the voltage signal $z(t)$ supplied by the oscillator to the phase-sensitive detector is at the same frequency and phase with the bridge working voltage. After phase-sensitive detection and low-pass filtering, the measured signal $\hat{x}(t)$ with the same polarity as the original can be obtained after amplification. This signal can propel the meter or connect to the subsequent instrument.

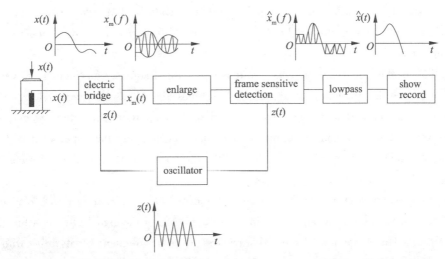

Figure 5-16  **Block diagram of dynamic resistance strain gauge in practical application**

## 5.3.2  Frequency modulation

### 5.3.2.1  Frequency modulation (FM) wave and its spectrum

The amplitude of the signal $x(t)$ is used to modulate the carrier frequency, called frequency modulation. In other words, the FM wave is a constant amplitude wave with different densities varying with the voltage amplitude of the signal $x(t)$, as shown in Figure 5-17.

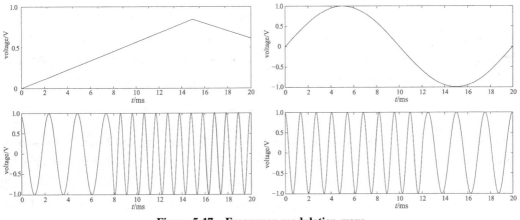

Figure 5-17  **Frequency modulation wave**

An important advantage of frequency modulation over amplitude modulation is that it improves signal-to-noise ratio. The analysis shows that in the case of amplitude modulation, if the interference noise and the carrier wave have the same frequency, the power ratio of the effective amplitude modulation wave to interference wave must be more than 35 dB.

However, in the case of frequency modulation, when the same performance index is reached in the case of amplitude modulation, the effective power ratio of frequency modulation to interference is only 6 dB. The reason why frequency modulation wave improves the signal-to-noise ratio in the signal transmission is that the information carried by FM signal is contained in the change of frequency, not in the amplitude, and the effect of interference wave is mainly reflected in the amplitude.

The FM method also has serious shortcomings: the frequency modulation wave usually requires a very wide frequency band, up to 20 times the bandwidth required by AM. The frequency modulation system is more complicated than the amplitude modulation system, because frequency modulation is a non-linear modulation that cannot apply the superposition principle. Therefore, the analysis of the FM wave is more difficult than the analysis of the AM wave. In fact, the analysis of the FM wave is approximate.

Frequency modulation is to make the frequency of the carrier signal vary with respect to the amplitude of the modulated signal $x(t)$. Since the amplitude of the signal $x(t)$ is a function that varies with time, the frequency of the FM wave should be a "frequency that varies with time". This seems to be difficult to understand. Next, we will analyze the real meaning of the above sentence in detail through the derivative relationship between angular frequency and time.

$$\omega = \frac{\mathrm{d}\phi}{\mathrm{d}t} \tag{5-26}$$

To be precise, $\omega$ should be the angular velocity, called the angular frequency (radians/sec, rad/s), or simply called the frequency, the phase angle in Equation (5-26) should be expressed as

$$u(t) = A\cos \phi \tag{5-27}$$

Usually, $\phi$ with a single frequency can be expressed as

$$\phi = \omega t + \theta \tag{5-28}$$

Where $\theta$ is the initial phase, which is a constant. So its derivative $\mathrm{d}\phi/\mathrm{d}t = \omega$, which is called the angular frequency. Instantaneous frequency $\mathrm{d}\phi/\mathrm{d}t$ is used to represent the modulation of the signal, which is frequency modulation, that is

$$\frac{\mathrm{d}\phi}{\mathrm{d}t} = \omega_0 [1 + x(t)] \tag{5-29}$$

Where $\omega_0$ is the center frequency of the carrier signal, $x(t)$ is the modulated signal, and $\omega_0 x(t)$ is the carrier signal modulated by the measured signal. This equation shows that the instantaneous frequency is the sum of the carrier center frequency $\omega_0$ and the frequency $\omega_0 x(t)$ that varies with the amplitude of the signal $x(t)$. By integrating both sides of Equation (5-29), the following results can be obtained

$$\phi = \omega_0 x(t) + \omega_0 \int x(t)\,\mathrm{d}t \tag{5-30}$$

If the modulation signal is the single cosine wave, that is

$$x(t) = A\cos(\omega t) \tag{5-31}$$

Then, the expression of the frequency modulation wave is

$$g(t) = G\sin \phi = G\sin\left[\omega_0 t + \omega_0\int x(t)\,\mathrm{d}t\right]$$

$$= G\sin\left[\omega_0 t + \omega_0\int A\cos \omega t\,\mathrm{d}t\right]$$

$$= G\sin\left[\omega_0 t + \frac{A\omega_0}{\omega}\sin \omega t\right] \tag{5-32}$$

$$= G\sin[\omega_0 t + m_f\sin \omega t]$$

Where $m_f = A\omega_0/\omega$ is called the frequency modulation index. $A\omega_0$ is the frequency amplitude of the actual change, which is called the maximum frequency bias, or expressed as $\Delta\omega = A\omega_0$. In order to study the frequency spectrum of FM wave, Equation (5-32) is expanded. At this time, by using Bessel function formula, the following equation can be obtained.

$$\cos(m_f\sin \omega t) = J_0(m_f) + 2\sum_{n=1}^{\infty} J_{2n}(m_f)\cos n\omega t \tag{5-33}$$

$$\sin(m_f\sin \omega t) = 2\sum_{n=1}^{\infty} J_{2n+1}(m_f)\sin[(2n+1)\omega t] \tag{5-34}$$

The following equation can be obtained by simple calculation

$$\begin{aligned}g(t) = G[&J_0(m_f)\sin \omega_0 t + J_1(m_f)\sin(\omega_0+\omega)t - \\ &J_1(m_f)\sin(\omega_0-\omega)t + J_2(m_f)\sin(\omega_0+2\omega)t + \\ &J_2(m_f)\sin(\omega_0-2\omega)t + \cdots + J_n(m_f)\sin(\omega_0+n\omega)t + \\ &(-1)^n J_n(m_f)\sin(\omega_0-n\omega)t]\end{aligned} \tag{5-35}$$

Where $J_n(m_f)$ is the Bessel function of order $n$ of $m_f$, $m_f$ is the independent variable, $n$ is an integer. Because of $m_f = A\omega_0/\omega$, it not only depends on the maximum frequency bias $\Delta\omega = A\omega_0$, but also depends on the modulation signal frequency $\omega$.

According to the above analysis, the frequency spectrum of the frequency modulation wave is obtained, as shown in Figure 5-18, and the following conclusions can be drawn from this:

(1) When expressed by a single frequency $\omega$, the frequency modulation wave can be expressed in the sum of the carrier frequency $\omega_0$ and many symmetrical side frequencies on both sides of the carrier frequency $(\omega_0+n\omega)$. The adjacent side frequencies are different from each other by $\omega$.

(2) The amplitude of each frequency component is equal to $GJ_n(m_f)$. When $n$ is even, the high and low symmetrical side frequencies have the same sign (positive or negative); when $n$ is an odd number, their signs (positive or negative) are opposite.

(3) Theoretically, the number of side frequencies is infinite. However, since starting from $n = m_f+1$, with the increase of $n$, the amplitude in side frequency decays quickly. It can be considered that the number of the effective side frequency is $2(m_f+1)$.

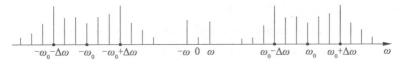

Figure 5-18　Frequency spectrum of FM wave

### 5.3.2.2 The application of frequency modulation in engineering test—DC frequency modulation and frequency discrimination

When using capacitance sensors, eddy current sensors or inductance sensors to measure displacement, force and other parameters, capacitance $C$ and inductance $L$ are often used as a tuning parameter of the resonant circuit of the self-excited oscillator. At this time, the resonant frequency of the oscillator is $\omega$

$$\omega = \frac{1}{\sqrt{LC}} \tag{5-36}$$

For example, when the capacitance $C$ is used as the tuning parameter in a capacitance sensor, differentiate the above equation, then

$$\frac{\partial \omega}{\partial C} = -\frac{1}{2}(LC)^{-\frac{3}{2}}L = \left(-\frac{1}{2}\right)\frac{\omega}{C} \tag{5-37}$$

Set $C = C_0$, then $\omega = \omega_0$, so the frequency increment is

$$\Delta\omega = \left(-\frac{1}{2}\right)\frac{\omega_0}{C_0}\Delta C$$

Therefore, when the parameter $C$ varies in a certain range, the instantaneous frequency of the resonance circuit is

$$\omega = \omega_0 \pm \Delta\omega = \omega_0\left(1 \mp \frac{\Delta C}{2C_0}\right) \tag{5-38}$$

It can be seen from Equation (5-38) that the oscillation frequency of the loop circuit has a linear relationship with the FM parameter, that is, within a certain range, it has a linear relationship with the change of the measured parameter. It is a frequency modulated mode where $\omega_0$ corresponds to the center frequency, and $\omega_0 \Delta C / 2C_0$ corresponds to the modulation part. This kind of circuit that converts the change of the measured parameter into the change of the oscillation frequency is called a direct frequency modulation measuring circuit.

The demodulation of FM wave, or frequency discrimination, is the process of converting frequency change into voltage amplitude change. In some test instruments, the transformer coupling resonant circuit is often used, as shown in Figure 5-19. In this figure, $L_1$ and $L_2$ are the primary and secondary coils of the transformer coupling, and they form a parallel resonant circuit with $C_1$ and $C_2$.

Input the constant-amplitude frequency modulation wave $e_f$. At the resonance frequency $f_n$ of the circuit, the coupling current in the coils $L_1$ and $L_2$ is the maximum, and the output voltage of the secondary side $e_a$ is also the maximum. When the frequency $e_f$ gets further away from $f_n$, $e_a$ and $e_f$ also decrease. Although the frequency of $e_a$ is consistent with that of $e_f$, the amplitude of $e_a$ varies with frequency, as shown in Figure 5-19b.

The Frequency-voltage variations are usually achieved by the subresonant region of the characteristic curve $e_a - f$, which approximates a straight line. When the measured parameter (such as displacement) is zero, the rising part of the characteristic curve corresponding to the oscillation frequency $f_o$ of the frequency modulation loop is approximately the midpoint of the straight line segment.

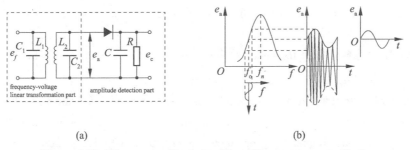

(a)                                                   (b)

**Figure 5-19   Frequency discrimination with resonance amplitude**

As the measured parameters change, the amplitude $e_a$ changes approximately linearly with the frequency of the FM wave, but the frequency of the FM wave $e_f$ remains an approximately linearly with the measured parameters. Therefore, by the amplitude detection of $e_a$, the change information of the measured parameter can be obtained, and the approximate linear relationship is maintained.

# Questions

5.1   A resistance wire strain gauge with resistance $R = 120\ \Omega$ and sensitivity $S = 2$ and a fixed resistance with a resistance value of $120\ \Omega$ form a bridge circuit. The supply voltage is 3 V and the load is assumed to be infinite. When the strain of the strain gauge is $2\ \mu\varepsilon$ and $2\ 000\ \mu\varepsilon$, calculate the output voltage of the single-arm and double-arm bridges respectively, and compare the sensitivity of the two cases.

5.2   Someone found that the sensitivity was insufficient when using resistance strain gauges, so they tried to increase the number of resistance strain gauges on the working bridge to improve sensitivity. Whether the sensitivity can be increased in the following situations? Why?

(1) One half bridge with two arms connected in series.

(2) The two arms of the half bridge are connected in parallel.

5.3   Connect a resistance strain gauge with a certain sensitivity of $S_g$ to form a full bridge circuit, and measure the strain of a certain component. It is known that the change pattern is $x(t) = A\cos 10t + B\cos 100t$, and if the bridge excitation voltage is $u_0 = E\sin 10\ 000t$. Find the frequency spectrum of the output signal of this bridge $U_y$.

5.4   What is the resolution of the filter? Related to those factors?

5.5   Suppose the lower cut-off frequency of a band-pass filter is $f_{c1}$, the upper cut-off frequency is $f_{c2}$, and the center frequency is $f_c$. Try to find out the correctness and errors in the following techniques.

（1）Frequency range filter $f_{c2} = \sqrt{2} f_{c1}$.

（2）$f_c = \sqrt{f_{c1} \cdot f_{c2}}$.

（3）The cut-off frequency of the filter is the frequency at the amplitude $-3$ dB of the passband.

（4）When the lower limit frequency is the same, the center frequency of the octave filter is $\sqrt[3]{2}$ times the center frequency of the 1/3 octave filter.

5.6    According to the response characteristics of the first-order $RC$ low-pass filter to the unit step input, illustrate the relationship between bandwidth $B$ and signal settling time $T_e$.

5.7    A signal has frequency components ranging from 100 Hz to 500 Hz. If amplitude modulation is performed on this signal, try to answer:

（1）What is the bandwidth of the amplitude-modulated wave?

（2）If the carrier frequency is 10 kHz, what frequency components will appear in the amplitude-modulated wave?

5.8    Can amplitude-modulated waves be regarded as the superposition of carrier and modulated signal? why?

5.9    Given the amplitude-modulated wave

$x(t) = (100 + 30\cos 2\pi f_1 t + 20\cos 6\pi f_1 t)(\cos 2\pi f_c t)$ , where $f_c = 10$ kHz, $f_1 = 500$ Hz, try to answer:

（1）The frequency and amplitude of each component included;

（2）Draw the frequency spectrum of the modulated signal and the amplitude-modulated wave.

5.10    Figure 5-20 is a block diagram of an AM demodulation system composed of a multiplier. Set the carrier signal be a sine wave with frequency $f_0$. Try to answer:

（1）Time-domain waveform of the output signal of each link;

（2）The frequency spectrum of the output signal of each link.

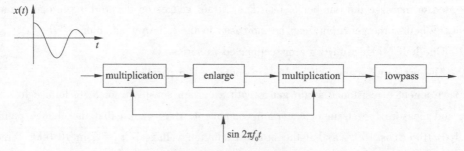

**Figure 5-20    Question 5.10**

# Chapter 6

## Signal Analysis and Processing

In the actual test, the measured signals are often mixed with various useless signals (collectively referred to as noise), and a lot of useful information is "overwhelmed". Generally, the source of noise is very complicated. It may be generated by the measured mechanical parts, or the system may have other input sources. Therefore, only after necessary analysis and processing of the measurement signal can the useful feature information be extracted. With the development of computer software and hardware technology, digital signal processing can already be implemented on general-purpose computers, providing good technical means for online machinery monitoring and real-time dynamic analysis.

In this chapter, the time domain processing method, the frequency domain processing method, the basis of digital signal processing and other related knowledge will be introduced in detail.

## 6.1 Signal processing method in time domain

In order to understand the properties of signals from time-domain waveform, the complex signal can be decomposed into several simple signals, and then the simple signals can be analyzed and processed in time domain.

Suppose the signal $x(t)$ can be decomposed into the sum of the steady-state component $x_d(t)$ and the alternating component $x_a(t)$, as shown in Figure 6-1. Namely

$$x(t) = x_d(t) + x_a(t) \tag{6-1}$$

The steady-state component is a quantity with regular changes, sometimes called trend quantity, while the alternating component contains the signal amplitude, frequency, phase and other information. Of course, it may also be random interference noise.

Mechanical testing technology

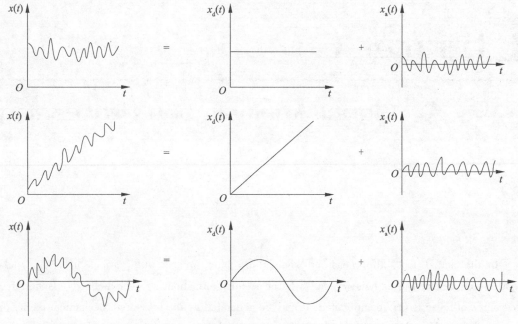

**Figure 6-1 The signal decomposed into the sum of the steady-state component and the
alternating component**

## 6.1.1 Statistical parameters in time-domain

Some statistical characteristic parameters can be obtained from the time-domain waveform,
which are often used for quick evaluation and simple diagnosis of mechanical equipment.

### 6.1.1.1 Dimensional amplitude parameter

The dimensional amplitude parameters include square root amplitude, average amplitude,
mean square amplitude and peak amplitude. If the signal $x(t)$ conforms to the stationary and
ergodic conditions of all states and its mean value is zero, set $x$ as the amplitude, and $p(x)$ as
the amplitude probability density function, and the dimensional amplitude parameter can be
defined as:

$$x_d = \left[ \int_{-\infty}^{+\infty} |x|^l p(x)\,dx \right]^{1/l} = \begin{cases} x_r, & l = 1/2 \\ \bar{x}, & l = 1 \\ x_{rms}, & l = 2 \\ x_p, & l \to \infty \end{cases} \tag{6-2}$$

Where $x_r$ is the root square amplitude; $\bar{x}$ is the mean value; $x_{rms}$ is the mean square value; $x_p$ is
the peak value. When $t \in (0,T)$, another definition of the above parameters is:

$$x_d = \begin{cases} x_r = \left[ \dfrac{1}{T} \int_0^T \sqrt{|x(t)|}\, dt \right]^2 \\[3mm] \overline{x} = \dfrac{1}{T} \int_0^T |x(t)|\, dt \\[3mm] x_{rms} = \left[ \dfrac{1}{T} \int_0^T x^2(t)\, dt \right]^{\frac{1}{2}} \\[3mm] x_p = E[\max |x(t)|] \end{cases} \tag{6-3}$$

The above four amplitude parameters are shown in Figure 6-2. Since the dimensional amplitude parameters are used to describe the mechanical state, they are not only related to the state of the machine, but also related to the motion parameters of the machine (such as speed, load, etc.). Therefore, it is impossible to draw a unified conclusion by directly using them to evaluate machines under different working conditions.

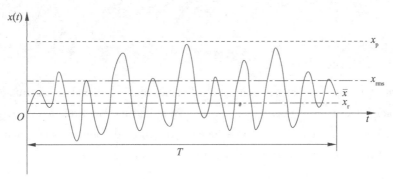

**Figure 6-2   Dimensional amplitude parameters**

## 6.1.1.2   Dimensionless parameters

Dimensionless parameters are insensitive to the variation of mechanical working conditions, which means that theoretically they are independent of the motion conditions of the machine and only depend on the shape of probability density function $p(x)$, therefore, dimensionless parameters are more suitable for evaluating the running state of mechanical equipment. Generally, it can be defined as

$$\zeta_x = \frac{\left[ \int_{-\infty}^{+\infty} |x|^l p(x)\, dx \right]^{1/l}}{\left[ \int_{-\infty}^{+\infty} |x|^m p(x)\, dx \right]^{1/m}} \tag{6-4}$$

According to the general definition of Formula (6-4), the following indicators can be obtained.

① Shape Factor $l=2$, $m=1$

$$K = \frac{x_{rms}}{\overline{x}}$$

② Crest Factor $l \to \infty$, $m=2$

$$c = \frac{x_p}{x_{rms}}$$

③ Impulse Factor $l \to \infty$, $m = 1$

$$I = \frac{x_p}{\bar{x}}$$

④ Clearance Factor $l \to \infty$, $m = 1/2$

$$L = \frac{x_p}{x_r}$$

In addition, you can also use higher-order statistics, such as the fourth moment $\alpha_4 = \int_{-\infty}^{+\infty} x^4(t) p(x) \, dx$ to define.

⑤ Kurtosis Value

$$K = \frac{\alpha_4}{\sigma_x^4}$$

Where $\sigma_x$ is the standard deviation, $\sigma_x = \left\{ \int_{-\infty}^{+\infty} [x(t) - \bar{X}]^2 p(x) \, dx \right\}^{\frac{1}{2}}$.

Although none of the above-mentioned dimensionless indicators constructed based on dimensional amplitude parameters are derived through strict functional relations or equations, they try to reflect the physical nature of machine state changes from different aspects, and can meet the requirements of those who are sensitive to machine state and requirements that are not sensitive to operating parameters.

For example, the kurtosis value is mainly used to detect impact components in mechanical signals. When the operating conditions of the machine change, such as the increase of speed and load, the signal amplitude and standard deviation also increase, but their ratio change is much smaller than that of the amplitude and standard deviation, so the kurtosis value becomes less sensitive to the changes of the operating conditions of the machine. On the other hand, when the signal has a shock pulse, the numerator in the kurtosis expression contains the fourth power factor of the signal amplitude, while the denominator is the second power factor of the amplitude. At this time, the kurtosis value will increase and deviate from the normal state. Therefore, kurtosis has a good diagnostic ability for signal impact sensitivity.

## 6.1.2 Correlation analysis

The so-called correlation refers to the linear relationship between variables. For deterministic signals, the relationship between two variables can be described by a function, and the two have the one-to-one correspondence and are a determined numerical relationship. There is no such definite relationship between two random variables. However, if there is some inherent physical connection between the two variables, then through a large number of statistics, it can be found that there is still some approximate relationship between them that characterizes their characteristics.

For example, in a gearbox, the relationship between the fatigue stress of the rolling bearing raceway and the axial load cannot be described by a deterministic function, however, through a large number of statistics, it can be found that the fatigue stress is correspondingly larger when the axial load is larger, and there is a certain linear relationship between these two variables.

For a random signal, the self-correlation function can be used to evaluate the correlation degree of its amplitude change at different time points. For two random signals, the corresponding cross-correlation functions can also be defined to represent the interdependence of their amplitude.

## 6.1.2.1　Correlation function

Set the time difference $\tau$ between the two signals, and then the correlation between the two signals in time shift can be studied. The correlation function is defined as

$$R_{xy}(\tau) = \int_{-\infty}^{\infty} x(t) y(t - \tau) \, dt \tag{6-5}$$

or
$$R_{yx}(\tau) = \int_{-\infty}^{\infty} y(t) x(t - \tau) \, dt$$

Obviously, the correlation function is a function of the time difference $\tau$. Usually $R_{xy}(\tau)$ or $R_{yx}(\tau)$ is called a cross-correlation function, If $x(t) = y(t)$, then $R_{xx}(\tau)$ or $R_x(\tau)$ is called the auto-correlation function, The above formula becomes

$$R_x(\tau) = \int_{-\infty}^{\infty} x(t) x(t - \tau) \, dt \tag{6-6}$$

If $x(t)$ and $y(t)$ are power limited signals, the correlation functions are defined as

$$R_{xy}(\tau) = \lim_{T \to \infty} \frac{1}{T} \int_{-\frac{T}{2}}^{\frac{T}{2}} x(t) y(t - \tau) \, dt \tag{6-7}$$

$$R_{yx}(\tau) = \lim_{T \to \infty} \frac{1}{T} \int_{-\frac{T}{2}}^{\frac{T}{2}} y(t) x(t - \tau) \, dt \tag{6-8}$$

$$R_x(\tau) = \lim_{T \to \infty} \frac{1}{T} \int_{-\frac{T}{2}}^{\frac{T}{2}} x(t) x(t - \tau) \, dt \tag{6-9}$$

From the above analysis, it can be seen that the correlation function of energy signal and power signal has different dimensions, the former is energy, and the latter is power.

## 6.1.2.2　The properties and application of self-correlation function

(1) The properties of the self-correlation function

According to the self-correlation function defined by Equation (6-6), the self-correlation function of a stationary random signal has nothing to do with $t$. The self-correlation function $R(\tau)$ mainly has the following properties:

① When $\tau = 0$, $R(\tau)$ has the maximum value and is equal to its mean square value;

② $R(\tau)$ is an even function, so it has $R(\tau) = R(-\tau)$, therefore, in practice we only need to get the $R(\tau)$ value at $\tau \geqslant 0$, not at $\tau < 0$;

③ When $\tau \neq 0$, the value of $R(\tau)$ is always less than that of $R_x(0)$;

④ A stationary signal with a mean value of zero. If $\tau \to +\infty$, $x(t)$ and $x(t+\tau)$ are not correlated, then $R(\tau) \to 0$;

⑤ If the stationary signal contains periodic components, its self-correlation function also contains periodic components, and its period is the same as that of the original signal. It can be proved that the self-correlation function of simple harmonic signal $x(t) = x_0 \sin(\omega_0 t + \varphi)$ is a cosine function.

$$R(\tau) = \frac{x_0^2}{2}\cos(\omega_0 \tau)$$

It is a periodic signal, and its period is the same as that of the original harmonic signal, but the phase information of the original signal is lost. The self-correlation function graphs of several common signals are shown in Figure 6-3. The self-correlation function of sinusoidal signal is shown in Figure 6-3b. For the narrow-band random signal shown in Figure 6-3e, the self-correlation function decays slowly (Figure 6-3f), while for the broadband random signal shown in Figure 6-3g, its self-correlation function will decay rapidly (Figure 6-3h); both the signals shown in Figure 6-3a and Figure 6-3c contain periodic components. It can also be seen from Figure 6-3b and Figure 6-3d that their corresponding self-correlation function curves do not decay to zero. In other words, the self-correlation function is an important way to find periodic signals or instantaneous signals from interference noise, that is, by extending the value of the variable $\tau$, the periodic components in the signal will appear gradually.

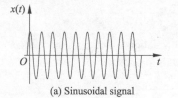

(a) Sinusoidal signal

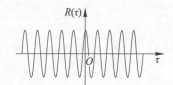

(b) Self-correlation function of sinusoidal signal

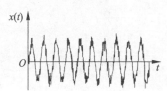

(c) Sinusoidal signal and random noise signal

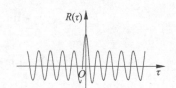

(d) Self-correlation function of sinusoidal signal and random noise signal

(e) Narrow-band random noise

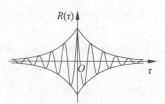

(f) Self-correlation function of narrow-band random noise

(g) Broadband random noise          (h) Self-correlation function of
                                          broadband random noise

**Figure 6-3   Several typical signals and their self-correlation functions**

（2）The application of self-correlation function

When using the sound signal to diagnose the running state of the machine, the sound signal of the machine in normal operation is composed of a large number of random impulse noises of nearly equal amplitude, so it has a wide and uniform spectrum. When the machine is running abnormally, there will be regular and periodic pulse signals in the random noise, the magnitude of which is much larger than the random impact noise.

For example, when the bearing in the mechanism wears and the gap increases, there will be an impact between the shaft and the bearing cover. Similarly, if the raceway of the rolling bearing is eroded and one of the meshing surfaces of the gear is seriously worn, periodic signals will appear in random noise. Therefore, when the machine fault is diagnosed through sound signal, the hidden periodic component can be found in the noise at first. Especially in the early stage of the fault, when the periodic signal is not obvious and it is difficult to find by direct observation, self-correlation analysis method can be used to detect the defect of the machine by relying on the amplitude of $R(\tau)$ and the frequency of fluctuation.

The self-correlation function of noise signal of the machine tool transmission is shown in Figure 6-4.

Figure 6-4a is the self-correlation function of noise under normal conditions. With the increase of $\tau$, $R(\tau)$ rapidly approaches the abscissa, indicating that the noise of gearbox is random noise. On the contrary, in Figure 6-4b, the self-correlation function $R(\tau)$ contains periodic components, when $\tau$ increases, $R(\tau)$ does not approach to the abscissa, which indicates that the gearbox is in abnormal working condition. Comparing the speed of each shaft in the gearbox with the fluctuation frequency of $R(\tau)$, the location of this defect can be determined.

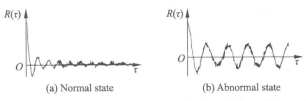

(a) Normal state          (b) Abnormal state

**Figure 6-4   Self-correlation function of acoustic signal of machine tool gearbox**

### 6.1.2.3   Cross-correlation function and its application

（1）The properties of cross-correlation functions

For two signals, a cross-correlation function can be used to characterize the interdependence between their amplitudes. Suppose two random signals are $x(t)$ and $y(t)$, then the cross-correlation function $R_{xy}(\tau)$ is

$$R_{xy}(\tau) = \lim_{T \to \infty} \frac{1}{T} \int_0^T x(t) y(t + \tau) \, \mathrm{d}t \qquad (6\text{-}10)$$

The cross-correlation function $R_{xy}(\tau)$ of stationary random signals is a real function, which can be positive or negative. Unlike an auto-correlation function, it is not an even function, and it does not have to be the maximum at $\tau = 0$. $R_{xy}(\tau)$ has the following main properties:

① Anti-symmetry, that is, $R_{xy}(-\tau) = R_{yx}(\tau)$;

② $[R_{xy}(\tau)]^2 \leqslant R_x(0) R_y(0)$;

③ For random signals $x(t)$ and $y(t)$, if there is no periodic component of the same frequency between them, then they are independent of each other when the time difference $\tau$ is large.

The graph of the cross-related functions $R_{xy}(\tau)$ shown in Figure 6-5 has a maximum value at $\tau = \tau_0$, it means that $x(t)$ and $y(t)$ have some connection in $\tau = \tau_0$, but not at other time difference. In other words, it reflects the lag time of the main transmission channel between $x(t)$ and $y(t)$. If the two signals have periodic components with the same frequency, even $\tau \to \infty$, it will have periodic components of that frequency;

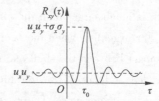

**Figure 6-5   Schematic diagram of cross-correlation function**

④ The cross-correlation function of two periodic signals with zero mean values and the same frequency retains the information of the circular frequency $\omega$, the corresponding amplitude $x_0$ and $y_0$ and the phase difference $\varphi$ of the two signals.

If two periodic signals are expressed as $x(t) = x_0 \sin(\omega t + \theta)$, $y(t) = y_0 \sin(\omega t + \theta - \varphi)$, between them, $\theta$ is the phase angle of $x(t)$ at $t = 0$, $\phi$ is the phase difference between $x(t)$ and $y(t)$, then the cross-correlation function of the two signals can be obtained as $R_{xy}(\tau) = \frac{1}{2} x_0 y_0 \cos(\omega t - \varphi)$.

(2) The application of cross-correlation function

The characteristics of cross-correlation functions are of great value in mechanical engineering applications. The following examples illustrate its application effects.

**Example 6-1**   Use the correlation analysis method to determine the location of the cracks of the deep buried oil pipeline for excavation and repair.

As shown in Figure 6-6, the leakage point $K$ can be regarded as the sound source that transmits sound to both sides. Sensor No.1 and sensor No.2 are placed on both sides of the pipe respectively. Because the two points where the sensors are placed are not at the same distance from the leakage point, the time for the sound of the oil leak to reach the two sensors will be different, on the cross-correlation function graph, there is a maximum at $\tau = \tau_m$, and this $\tau_m$ is the time difference. Suppose $S$ is the distance between the installation center line of the two sensors and the leakage point, and $\nu$ is the propagation speed of sound in the pipeline, then:

$$S = \frac{1}{2}\nu\tau_m$$

Use $\tau_m$ to determine the location of the leak, which is a linear positioning problem. The positioning error is tens of centimeters. This method can also be used for curved pipes.

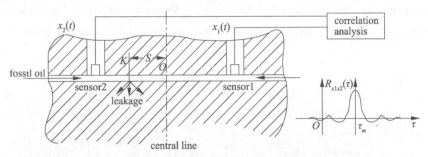

**Figure 6-6   Examples of linear positioning using correlation analysis**

**Example 6-2**   Use the correlation method to test the speed of the hot-rolled steel strip.

Figure 6-7 shows an example of using cross-correlation analysis to measure the velocity of hot-rolled steel strip online. Install two convex lenses and two photocells in the direction of steel plate moving. The distance between two convex lenses is $d$. When the hot-rolled steel strip moves at speed $v$, the light reflected from the surface of the hot-rolled steel strip is focused on two photovoltaic cells at a distance of $d$ through the lens. The fluctuation of the reflected light intensity is converted into an electrical signal by photovoltaic cells. Then perform cross-correlation analysis for these two electrical signals, and measure the time $\tau_m$ corresponding to the maximum value of the cross-correlation function through the adjustable delay device. Since the signals $x(t)$ and $y(t)$ generated when any section $P$ of the steel strip passes through points $A$ and $B$ are completely correlated, the maximum value can be generated on the cross-correlation curve, and the moving speed of the hot-rolled steel strip is $v = d/\tau_m$.

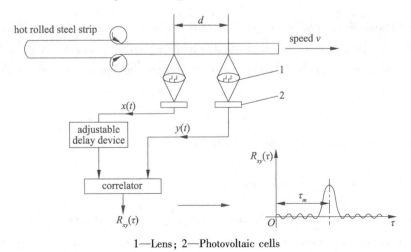

1—Lens; 2—Photovoltaic cells

**Figure 6-7   Use correlation analysis to measure speed**

**Example 6-3** Use cross-correlation function for equipment fault diagnosis.

Check whether the vibration of the driver's seat of the car is caused by the engine or the rear axle. Acceleration sensors can be arranged on the engine, driver's seat, and rear axle, as shown in Figure 6-8. Then amplify the output signal and perform correlation analysis. It can be seen that the correlation between the engine and the driver's seat is poor, while the cross-correlation between the rear axle and the driver's seat is large. Therefore, it can be considered that the vibration of the driver's seat is mainly caused by the vibration of the rear axle of the car.

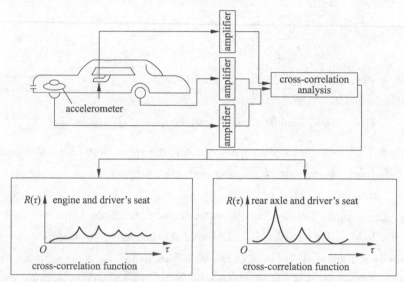

Figure 6-8 Identification of vehicle vibration transmission path

# 6.2 Signal processing method in frequency domain

In mechanical signal processing, the Fourier transform regards a random signal as the synthesis of multiple sinusoidal waves with different frequencies, which makes it possible to analyze the random signal in the frequency domain. Mechanical signals processing methods in the frequency domain, including self-power spectrum, cross-power spectrum, coherence analysis, etc. are introduced in this section.

## 6.2.1 Self power spectral density function

### 6.2.1.1 Definition and its physical meaning

Suppose that $x(t)$ is a random signal with zero mean value, that is, $\mu_x = 0$ (if the original random signal is non-zero mean value, it can be properly processed to make the mean value zero), and that there is no periodic component in $x(t)$, then when $\tau \to \infty$, $R_x(\tau) \to 0$. In this way, the self-correlation function $R_x(\tau)$ can satisfy the condition $\int_{-\infty}^{\infty}|R_x(\tau)|d\tau < \infty$ for the Fourier transform. Then it can get the Fourier transform $S_x(f)$ of $R_x(\tau)$.

$$S_x(f) = \int_{-\infty}^{\infty} R_x(\tau)\,e^{-j2\pi ft}d\tau \qquad (6\text{-}11)$$

Inverse transformation

$$R_x(\tau) = \int_{-\infty}^{\infty} S_x(f)\,e^{j2\pi ft}df \qquad (6\text{-}12)$$

Define $S_x(f)$ as the self-power spectral density function of $x(t)$, which is referred to as self-spectrum or self-power spectrum. Since the relationship between $S_x(f)$ and $R_x(\tau)$ is a Fourier transform pair, the two are uniquely corresponding. Because $R_x(\tau)$ is a real even function, $S_x(f)$ is also a real even function. Therefore, $G_x(f) = 2S_x(f)$ in the range of $f = (0 \sim \infty)$ is often used to represent the entire power spectrum of the signal, and $G_x(f)$ is called the unilateral power spectrum of the signal $x(t)$, as shown in Figure 6-9.

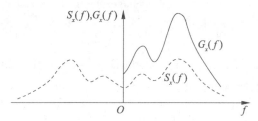

**Figure 6-9   Unilateral and bilateral spectra**

If $\tau = 0$, according to the definition of self-correlation function $R_x(\tau)$ and auto-power spectral density function $S_x(f)$, we can get

$$R_x(0) = \lim_{T\to\infty} \frac{1}{T}\int_0^T x^2(t)\,dt = \int_{-\infty}^{\infty} S_x(f)\,df \qquad (6\text{-}13)$$

Thus, the area under the curve $S_x(f)$ and surrounded by the frequency axis is the average power of the signal, and $S_x(f)$ is the distribution of the power density of signal along the frequency axis, so $S_x(f)$ is called the self-power spectral density function.

## 6.2.1.2   Pasiphae theorem

The total energy of the signal calculated in the time domain is equal to the total energy of the signal calculated in the frequency domain. This is the Pasiphae theorem, which is

$$\int_{-\infty}^{\infty} x^2(t)\,dt = \int_{-\infty}^{\infty} |X(f)|^2 df \qquad (6\text{-}14)$$

Equation (6-14) is also called the energy equation. This theorem can be derived from the convolution equation of Fourier transform.

Suppose:

$$\begin{aligned} x(t) &\Leftrightarrow X(f) \\ h(t) &\Leftrightarrow H(f) \end{aligned} \qquad (6\text{-}15)$$

According to the convolution theorem in the frequency domain,

$$x(t)h(t) \Leftrightarrow X(f) * H(f) \qquad (6\text{-}16)$$

which is

$$\int_{-\infty}^{\infty} x(t)h(t)\,e^{-j2\pi qt}\,dt = \int_{-\infty}^{\infty} X(f)H(q-f)\,df \qquad (6\text{-}17)$$

Set $q=0$, get

$$\int_{-\infty}^{\infty} x(t)h(t)\,dt = \int_{-\infty}^{\infty} X(f)H(-f)\,df \qquad (6\text{-}18)$$

Set $h(t)=x(t)$, get

$$\int_{-\infty}^{\infty} x^2(t)\,dt = \int_{-\infty}^{\infty} X(f)X(-f)\,df \qquad (6\text{-}19)$$

$x(t)$ is a real function, then $X(-f)=X^*(f)$, so

$$\int_{-\infty}^{\infty} x^2(t)\,dt = \int_{-\infty}^{\infty} X(f)X^*(f)\,df = \int_{-\infty}^{\infty} |X(f)|^2\,df \qquad (6\text{-}20)$$

$|X(f)|^2$ is called the energy spectrum, which is the energy distribution density along the frequency axis. On the entire time axis, the average power of the signal is

$$P_{av} = \lim_{T \to \infty} \frac{1}{T} \int_0^T x^2(t)\,dt = \int_{-\infty}^{+\infty} \lim_{T \to \infty} \frac{1}{T} |X(f)|^2 df \qquad (6\text{-}21)$$

According to Equation (6-11), the relationship between the auto-power spectral density function and the amplitude spectrum is

$$S_x(f) = \lim_{T \to \infty} \frac{1}{T} |X(f)|^2 \qquad (6\text{-}22)$$

Based on this relationship, the power spectrum can be calculated by directly performing Fourier transform on the time domain signal.

### 6.2.1.3 Application

The auto-power spectral density $S_x(f)$ is the Fourier transform of the self-correlation function $R_x(\tau)$, so $S_x(f)$ contains all the information of $R_x(\tau)$.

The auto-power spectral density $S_x(f)$ reflects the frequency domain structure of the signal, which is consistent with the amplitude spectrum $|X(f)|$, but the auto-power spectral density reflects the square of the signal amplitude, so the characteristics of its frequency domain structure are more obvious, as shown in Figure 6-10.

For a linear system (as shown in Figure 6-11), if its input is $x(t)$ and its output is $y(t)$, the frequency response function of the system is $H(f)$, $x(t) \Leftrightarrow X(f)$, $y(t) \Leftrightarrow Y(f)$, then

$$Y(f) = H(f)X(f) \qquad (6\text{-}23)$$

It is not difficult to prove that the relationship among the auto-power spectral density of input, the auto-power spectral density of output and the system frequency response function is as follows

$$S_y(f) = |H(f)|^2 S_x(f) \qquad (6\text{-}24)$$

Through the analysis for the input auto-spectrum and output auto-spectrum, the amplitude-frequency characteristics of the system can be obtained. However, the phase information is lost in such calculation, so the phase frequency information of the system cannot be obtained in this process.

　　Self-correlation analysis can effectively detect whether there are periodic components in the signal. Auto-power spectral density can also be used to detect periodic components of the signal. The frequency spectrum of a periodic signal is an impulse function, and the energy at a certain frequency is unlimited. But in actual processing, the signal is truncated with a rectangular window function, this is equivalent to performing convolution for the spectral function *sinc* (the spectral of the rectangular window function) and the spectral function $\delta$ (the spectral of the periodic signal) in the frequency domain. Therefore, the frequency spectrum of the truncated periodic signal is no longer an impulse function. The height of the original spectral line becomes finite, and the width of the spectral line changes from infinitesimal to a certain width. Therefore, the measured periodic components appear in the form of steep finite peaks in the power spectral density graph.

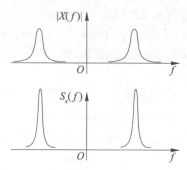

**Figure 6-10　Amplitude spectrum and auto-power spectrum**

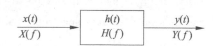

**Figure 6-11　Ideal single input, single output system**

## 6.2.2　Cross-spectral density function

### 6.2.2.1　Definition

　　If the cross-correlation function $R_{xy}(\tau)$ satisfies the Fourier transform condition $\int_{-\infty}^{\infty} |R_{xy}(\tau)| d\tau < 0$, then

$$S_{xy}(f) = \int_{-\infty}^{\infty} R_{xy}(\tau) e^{-j2\pi ft} d\tau \qquad (6\text{-}25)$$

$S_{xy}(f)$ is called the cross-spectral density function of signals $x(t)$ and $y(t)$, referred to as cross-spectrum for short. According to the Fourier transform, then

$$R_{xy}(\tau) = \int_{-\infty}^{\infty} S_{xy}(f) e^{j2\pi ft} df \qquad (6\text{-}26)$$

　　The cross-correlation function $R_{xy}(\tau)$ is not even, so $S_{xy}(\tau)$ has both virtual and real parts. Similarly, $S_{xy}(f)$ retains all the information in $R_{xy}(\tau)$.

### 6.2.2.2　Application

　　For the linear system shown in Figure 6-11, it can be proved that

$$S_{xy}(f) = H(f)S_x(f) \qquad (6\text{-}27)$$

Therefore, based on the auto-spectrum $S_x(f)$ and the cross-spectrum $S_{xy}(f)$, the frequency response function of the system can be obtained according to Equation (6-27). Equation(6-27) is different from Formula (6-24), the obtained $H(f)$ contains not only the amplitude-frequency characteristics but also the phase-frequency characteristics. This is because the cross-correlation function contains phase information.

If the test system is subject to the external interference, as shown in Figure 6-12, $n_1(t)$ is the input noise, $n_2(t)$ is the noise added to the middle part of the system, and $n_3(t)$ is the noise added to the output part. Obviously, the output $y(t)$ is as follows

$$y(t) = x'(t) + n'_1(t) + n'_2(t) + n'_3(t) \qquad (6\text{-}28)$$

Where $x'(t)$, $n'_1(t)$, $n'_2(t)$ are the response of the system to $x(t)$, $n_1(t)$ and $n_2(t)$ respectively.

The cross-correlation function of input $x(t)$ and output $y(t)$ is

$$R_{xy}(\tau) = R'_{xx}(\tau) + R'_{xn1}(\tau) + R'_{xn2}(\tau) + R'_{xn3}(\tau) \qquad (6\text{-}29)$$

Since the input $x(t)$ and the noise $n_1(t)$, $n_2(t)$, $n_3(t)$ are independent and unrelated, the cross-correlation functions $R'_{xn1}(\tau)$, $R'_{xn2}(\tau)$ and $R'_{xn3}(\tau)$ are all zero. So

$$R_{xy}(\tau) = R'_{xx}(\tau) \qquad (6\text{-}30)$$

Therefore
$$S_{xy}(\tau) = S'_{xx}(f) = H(f)S_x(f) \qquad (6\text{-}31)$$

where $H(f) = H_1(f)H_2(f)$ is the frequency response function of the system.

It can be seen that cross-spectrum analysis can eliminate the influence of noise. This is the outstanding advantage of this analytical method. However, it should be noted that when using Equation (6-31) to calculate linear system $H(f)$, although the cross-spectrum $S_{xy}(f)$ is not affected by noise, the auto-spectrum $S_x(f)$ of the input signal still cannot exclude the influence of the measurement noise at the input, thus forming a measurement error.

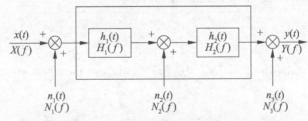

**Figure 6-12　The system subject to external interference**

In order to test the dynamic characteristics of the system, the specific input disturbances $z(t)$ are sometimes imposed on the running test system. It can be seen from Equation (6-31) that as long as $z(t)$ is independent of other inputs, $H(f)$ can be calculated after measuring $S_{xy}(f)$ and $S_z(f)$. This kind of test is called "on-line test" when the tested system is running normally.

### 6.2.3　Coherence function

It is important in many cases to evaluate the causality between the input signal and the output signal, that is, how much of the power spectrum of the output signal is the response caused by the input. This causation is usually described by the coherence function $r_{xy}^2(f)$, which is defined as:

$$r_{xy}^2(f) = \frac{|S_{xy}(f)|^2}{S_x(f)S_y(f)} \quad (0 \leqslant r_{xy}^2(f) \leqslant 1) \tag{6-32}$$

In fact, when using Equation (6-32) to calculate the coherence function, only the estimated values of $S_y(f)$, $S_x(f)$ and $S_{xy}(f)$ can be used, and the obtained coherence function is only an estimated value; and only by using $\hat{S}_y(f)$, $\hat{S}_x(f)$ and $\hat{S}_{xy}(f)$ after multi-stage smoothing, the obtained $\hat{r}_{xy}^2(f)$ is a better estimate.

If the coherence function is zero, the output signal is irrelevant to the input signal. When the coherence function is 1, it means that the output signal is completely coherent with the input signal, the system is not interfered and the system is linear. The value of the coherence function is in the range of $[0, 1]$, indicating that there are three possibilities: ① there is external noise interference during the test; ② the output $y(t)$ is the comprehensive output of input $x(t)$ and other inputs; ③ the system connecting $y(t)$ and $x(t)$ is nonlinear.

**Example 6-3**　The coherence analysis between pressure pipe vibration and pressure pulse of marine diesel oil pump is shown in Figure 6-13.

The speed of lubricating oil pump is $n=781$ r/min, the teeth number of the oil pump gear is $z=14$. Measured oil pressure pulsation signal is $x(t)$ and pressure oil pipe vibration signal is $y(t)$. The fundamental frequency of the pressure pulsation of the pressure pipe is $f_0=nz/60=182.24$ Hz.

In Figure 6-13c, when $f=f_0=182.24$ Hz, then $r_{xy}^2(f) \approx 0.9$; at $f=2f_0 \approx 361.12$ Hz, $r_{xy}^2(f) \approx 0.37$; at $f=3f_0 \approx 546.54$ Hz, $r_{xy}^2(f) \approx 0.8$; at $f=4f_0 \approx 722.24$ Hz, $r_{xy}^2(f) \approx 0.75 \cdots$ The coherence function values corresponding to each harmonic frequency caused by the gear are relatively large, while the coherence function values corresponding to other frequencies are small. It can be seen that the vibration of the oil pipe is mainly caused by the oil pressure pulsation. The influence of oil pressure pulsation can also be evident from the auto-spectra of $x(t)$ and $y(t)$ (see Figure 6-13a, b).

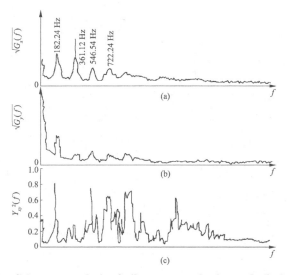

**Figure 6-13　Coherence analysis of oil pressure pulsation and oil pipe vibration**

# 6.3 Digital signal processing

Digital signal processing is very complicated, involving system analysis, sensor and its characteristics analysis, signal sampling, etc. In general, the working principle of digital signal processing is shown in Figure 6-14.

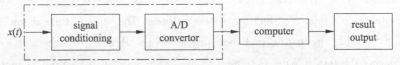

**Figure 6-14    Diagram of digital signal processing principle**

When using a digital analyzer or computer to analyze and process the signal, it is necessary to digitize the continuously measured dynamic signal and convert it into a discrete digital sequence. The digitization process mainly includes discrete sampling and amplitude quantization and sampling signal coding. First, the sample-and-hold amplifier samples the preprocessed analog signal into a discrete sequence according to the selected sampling interval, the signal at this time becomes a sampled signal with discrete time and continuous amplitude. Then, the quantization and coding device converts the amplitude of each sampled signal into a digital code, and finally turns the sampled signal into a digital sequence.

## 6.3.1    Signal sampling and sampling theorems

Signal sampling is the process of discretizing for a continuous signal. The sampling process is shown in Figure 6-15. Among them, the analog signal is $x(t)$, and the sampling pulse function with a sampling period of $T$ is $p(t)$.

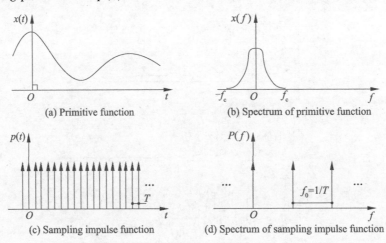

(a) Primitive function

(b) Spectrum of primitive function

(c) Sampling impulse function

(d) Spectrum of sampling impulse function

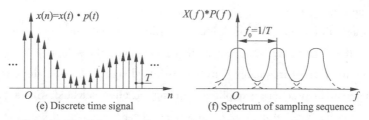

(e) Discrete time signal  (f) Spectrum of sampling sequence

**Figure 6-15   Schematic diagram of time domain sampling process**

The discrete time signal $x(n)$ can be obtained by multiplying $x(t)$ and $p(t)$, so

$$x(n) = \sum_{n=-\infty}^{\infty} x(nT)\,\delta(t-nT) \tag{6-33}$$

where $x(nT)$ is the value of the analog signal at $t=nT$. Suppose the Fourier transform of $x(t)$ is $X(f)$, and the Fourier transform $P(f)$ of pulse function $p(t)$ is also a pulse sequence, and the pulse interval is $1/T$, expressed as

$$P(f) = \frac{1}{T}\sum_{m=-\infty}^{\infty} \delta\!\left(f-\frac{m}{T}\right) \tag{6-34}$$

According to the convolution theorem in frequency domain: the Fourier transform of the product of two time-domain functions is equal to the convolution of the two Fourier transforms, then the Fourier transform $X(e^{j2\pi f})$ of the discrete sequence $x(n)$ can be written as

$$X(e^{j2\pi f}) = \frac{1}{T}\sum_{m=-\infty}^{\infty} X\!\left(f-\frac{m}{T}\right) \tag{6-35}$$

Equation (6-35) is the frequency spectrum of the sampled signal formed by $x(t)$ after sampling with a time interval of $T$. Generally speaking, this spectrum and the spectrum $X(f)$ of the original continuous signal are not necessarily the same, but they are related. It shifts the frequency spectrum $X(f)$ of the original signal by $1/T$ to the frequency domain sequence point corresponding to each sampling pulse, then all are superimposed, as shown in Figure 6-15. It can be seen that the signal becomes a discrete signal after time-domain sampling, and the frequency-domain function of the new signal becomes a periodic function, with a period of $f_s = 1/T$.

If the sampling interval $T$ is too large, that is, the sampling frequency $f_s$ is too small, so that the translation distance $1/T$ is too small, then a part of the spectrum $X(f)$ moved to each sampling pulse will overlap each other, and the newly synthesized $X(f)*P(f)$ pattern is inconsistent with the original $X(f)$. This phenomenon is called aliasing. When the aliasing occurs, parts of the amplitude of the original spectrum (the dotted portion in Figure 6-15f) are changed, so that it is impossible to accurately recover the original time domain signal $x(t)$ from discrete signal $X(n)$.

If $x(t)$ is a band-limited signal, that is, the highest frequency $f_c$ of the signal is a finite value (as shown in Figure 6-15b). When the sampling frequency is $f_s > 2f_c$, the spectrum after sampling will not be aliased. If the spectrum is passed through an ideal low-pass filter with a center frequency of zero and a bandwidth of $\pm f_s/2$, the spectrum of the original signal can be extracted, which means that it is possible to accurately restore the original analog signal $x(t)$ from the discrete sequence.

From the above discussion we can know, in order to avoid aliasing, it is still possible to accurately reflect the original signal after sampling processing, the sampling frequency $f_s$ must be greater than twice the highest frequency $f_c$ in the processed signal, that is, there is $f_s>2f_c$. This is the sampling theorem. In practice, sampling frequency is often selected with a margin of error, and should generally be selected as 3 to 4 times the highest frequency in the processed signal. In addition, if we can determine that the high frequency part of the measured signal is caused by interference noise, in order to meet the sampling theorem and not make the sampling frequency too high, we can first perform low-pass filtering on the measured signal.

## 6.3.2 Signal quantization and quantization error

The number of bits of the analog-to-digital converter is fixed, and it can only express levels with a certain interval. When the level of the analog signal sampling point falls between two adjacent levels, it must be rounded to a similar level. This process is called signal quantization. Assuming that the increment between two adjacent levels is $\Delta$, then the maximum value of the quantization error $\varepsilon$ is $\pm\Delta/2$. Moreover, it can be considered that the probability of $\varepsilon$ appearing in the interval $(-\Delta/2,\Delta/2)$ is equal, the probability distribution density is $1/\Delta$ and the mean value is zero, then the mean square value is

$$\sigma_z^2 = \int_{-\Delta/2}^{\Delta/2} \varepsilon^2 \frac{1}{\Delta}d\varepsilon = \frac{\Delta^2}{12} \qquad (6\text{-}36)$$

If the number of bits of the analog-to-digital converter is set to $N$, binary coding is used, and the conversion range of the converter is $\pm V$, then the increment $\Delta$ between adjacent levels can be expressed as

$$\Delta = \frac{V}{2^{N-1}} \qquad (6\text{-}37)$$

According to the discussion of quantization error and Equation (6-37), for the analog-digital conversion module with $N$-bit binary, the relative quantization error $\delta$ in the actual full range is

$$\delta = \frac{1}{2^{N-1}}\times 100\% \qquad (6\text{-}38)$$

The quantization error is the random error superimposed on the sampled signal, but in order to simplify the discussion of subsequent questions, it considers that the number of bits of the analog-to-digital converter is infinite, so that the amplitude collected by the sampling point is the amplitude on the original analog signal.

## 6.3.3 Signal truncation, signal energy leakage and window functions

Generally, the length of the signal is infinite, and it is impossible for us to process the entire signal of infinite length, so we have to truncate it. The truncation is the multiplication of an infinitely long signal by a function of finite width in the time domain. This function is called a window function. The simplest window function is a rectangular window, as shown in Figure 6-16.

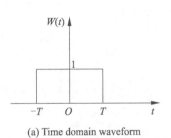

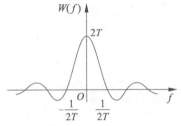

(a) Time domain waveform   (b) Amplitude-frequency characteristic curve

**Figure 6-16   Rectangular window function**

The rectangular window function $w(t)$ and its amplitude-frequency characteristic $W(f)$ can be expressed as

$$w(t) = \begin{cases} 1 & |t| < T \\ 0.5 & |t| = T \\ 0 & |t| > T \end{cases} \tag{6-39}$$

$$W(f) = 2T\frac{\sin(2\pi fT)}{2\pi fT} \tag{6-40}$$

If the original signal is $x(t)$, its spectrum function is $X(f)$, according to the convolution theorem in frequency domain, it can be known that the spectrum of the signal truncated by the rectangular window function is the convolution of $X(f)$ and $W(f)$. Since $W(f)$ is an infinite frequency band function, even if $x(t)$ is a limited band signal, the truncated spectrum must be a function of infinite bandwidth, indicating that the energy distribution of the signal has expanded; and since the truncated signal is an infinite bandwidth signal, no matter how high the sampling frequency is selected, mixing frequency will inevitably occur. It can be seen that signal truncation will inevitably lead to certain errors, and this phenomenon is called signal energy leakage.

If the truncation length is increased, that is, $T$ in Figure 6-17a increases, it can be seen from Figure 6-17b that the $W(f)$ graph will be compressed and narrowed. Although in theory the spectral range is still infinitely wide, the decay rate of frequency components other than the center frequency is accelerated, so the leakage error will be reduced. And when $T \rightarrow \infty$, the $W(f)$ function will become the $\delta(f)$ function, and the convolution of $W(f)$ and $\delta(f)$ is still $W(f)$, which means that there is no leakage error without truncation. In addition, signal energy leakage is also related to the side lobes of the window function spectrum. If the side lobe of the window function is small, the corresponding leakage is also small. In addition to rectangular windows, window functions commonly used in mechanical signal measurement include triangular windows and Hanning windows, as shown in Figure 6-17.

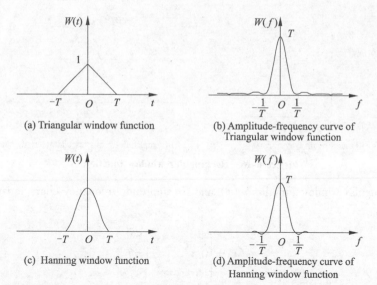

(a) Triangular window function

(b) Amplitude-frequency curve of Triangular window function

(c) Hanning window function

(d) Amplitude-frequency curve of Hanning window function

**Figure 6-17  Rectangular windows, Hanning windows and their amplitude-frequency characteristics**

The triangular window function $w(t)$ and its amplitude-frequency characteristic $W(f)$ are respectively expressed as

$$w(\tau)=\begin{cases}1-\dfrac{1}{T}|\tau| & |\tau|<T\\[2mm] 0 & |\tau|\geqslant T\end{cases} \tag{6-41}$$

$$W(f)=T\left[\frac{\sin(\pi fT)}{\pi fT}\right]^{2} \tag{6-42}$$

The Hanning window function $w(t)$ and its amplitude-frequency characteristic $W(f)$ are expressed as

$$w(\tau)=\begin{cases}\dfrac{1}{2}+\dfrac{1}{2}\cos\left(\dfrac{\pi\tau}{T}\right) & |\tau|<T\\[2mm] 0 & |\tau|\geqslant T\end{cases} \tag{6-43}$$

$$W(f)=\frac{1}{2}Q(f)+\frac{1}{4}\left[Q\left(f+\frac{1}{2T}\right)+Q\left(f-\frac{1}{2T}\right)\right] \tag{6-44}$$

$$Q(f)=\frac{\sin(2\pi fT)}{\pi fT}$$

The side lobe of the two window functions is discussed above. In particular, the side lobe of the Hanning window is much smaller than that of the rectangular window, so it can suppress the leakage error better. In the actual mechanical signal processing, the signal is often truncated by the unilateral window function, which is equivalent to time-shifting the bilateral window function, the time shift in the time domain corresponds to a phase shift in the frequency domain and the absolute value of the amplitude does not change, so the conclusion of leakage error caused by unilateral window function truncation is the same as above.

This book mainly introduces basic knowledge about signal processing method in time domain and frequency domain, and digital signal processing method. With the rapid development of modern industry and the continuous presentation of various application requirements, signal analysis and processing technology will continue to progress in signal processing speed, resolution, functional range and special processing, and new processing technologies will continue to emerge. At present, the development of signal processing is mainly reflected in the new signal processing technology, new signal processing methods and the real-time ability of signal processing. Limited by space, please refer to relevant books for details of modern signal analysis and processing methods, including power spectrum estimation, time-frequency analysis, statistical signal processing, etc.

# Questions

6.1   Find the self-correlation function of $h(t)$.

$$h(t) = \begin{cases} e^{-at} & (t \geqslant 0, a > 0) \\ 0 & (t < 0) \end{cases}$$

6.2   Find the cross-correlation function of the square wave and the sinusoidal wave (as shown in Figure 6-18).

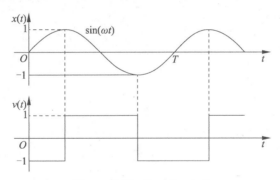

**Figure 6-18   Question 6.2**

6.3   Figure 6-19 shows the graph of correlation function of a measured signal. Try to analyze, is the graph a self-correlation function $R_x(\tau)$ graph or a cross-correlation function $R_{xy}(\tau)$ graph? Why? What information about the signal can be obtained from it?

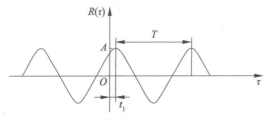

**Figure 6-19   Question 6.3**

6.4   What is frequency mixing phenomenon? What is the sampling theorem? How can we avoid spectrum aliasing?

6.5   Briefly explain the main function of each component link of the computer test system.

6.6   Give examples to illustrate the difference between smart sensors and classic sensors.

6.7   What is fieldbus technology? What are the advantages? Please illustrate with examples.

6.8    What is a virtual instrument? Compared with traditional instruments, what are the advantages of virtual instruments? Please illustrate with examples.

# Chapter 7

## The Application of Mechanical Testing Technology

This chapter mainly introduces the testing systems, testing method and application of stress strain, vibration, displacement and other mechanical quantities in mechanical engineering, aiming to help students better master the basic knowledge and application method of testing technology, and lay a good foundation for further study and scientific research.

## 7.1 Stress-strain test and its application

Stress-strain test is a common method to test, analyze, and evaluate the reliability and safety of engineering structure design, manufacture and assembly. Its main function is to test the residual stress on the workpiece surface, to analyze the magnitude and distribution of the residual stress on the workpiece surface, quantitatively to evaluate the influence of the residual stress on the fatigue strength, stress corrosion, dimensional stability, and service life of the workpiece, and to evaluate the heat treatment process, surface strengthening process and stress relieving process, etc.

### 7.1.1 Stress-strain test

#### 7.1.1.1 Strain sensor

The basic purpose of strain gauges is to measure strain, but it is far from limited to this in measurement. Any physical quantity that can be transformed into strain change can be measured with strain gauges. The key point is how to choose a suitable elastic element to convert the measured physical quantity into the change of strain. Strain gauges and elastic elements are two indispensable key parts of various strain sensors.

The elastic element structure of common strain sensors is shown in Figure 7-1. Figure 7-1a shows a diaphragm pressure strain sensor. Figure 7-1b shows a cylindrical force-strain sensor. Figure 7-1c shows a ring force-strain sensor. Figure 7-1d shows a torque-strain sensor. Figure 7-1e shows an octagonal ring turning dynamometer, which can be used to measure three mutually perpendicular forces at the same time ( Feeding resistance $F_x$, resistance to cutting $F_y$ and main cutting force $F_z$). Figure 7-1f shows an elastic beam strain accelerometer which also can be used in strain sensors for measuring displacements and forces.

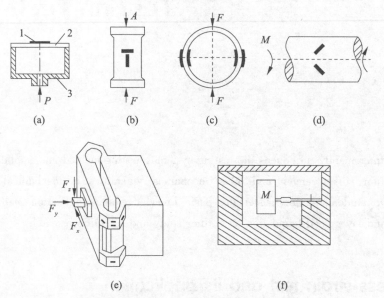

1—Strain gauge; 2—Diaphragm; 3—Case body

**Figure 7-1　Schematic diagram of the strain sensor**

## 7.1.1.2 Arrangement and bridge connection of strain gauges in various load measurements

Strain gauges sense the tension or compression deformation at a certain point on the surface of a component, and sometimes the strain may be caused by various internal forces ( such as tension and bending). In order to measure the deformation caused by a certain internal force and eliminate the strain of other internal forces, the patch position, direction and bridge construction mode must be reasonably selected, so as to make use of the addition and subtraction characteristics of the bridge to achieve the purpose of measurement and the effect of temperature compensation.

There are two methods to form the measuring bridge: the half-bridge circuit method and the full bridge circuit method. The half-bridge circuit refers to a bridge composed of two resistance strain gauges as adjacent arms of the bridge, and the other two arms are precision non-inductive resistors in the bridge box of the strain gauge. The full-bridge circuit means that all four legs of the bridge are composed of resistance strain gauges. This connection can eliminate the influence of connecting wire resistance and reduce the influence of contact resistance and improve sensitivity.

When carrying out load measurement, half-bridge measurement or full-bridge measurement can be adopted as required. During half-bridge measurement, the working half-bridge is connected with the terminals 1, 2 and 3 of the bridge box as shown in Figure 7-2a, and connected with the precise non-inductive resistor in the bridge box through the short connector to form a measuring circuit and access the strain gauge. When measuring the full-bridge, the full-bridge composed of working strain gauges connected with the terminals 1, 2, 3 and 4 of the bridge box as shown in Figure 7-2b, and the measuring circuit is connected to the strain gauge through the bridge box.

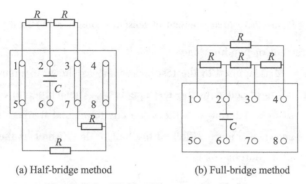

(a) Half-bridge method          (b) Full-bridge method

**Figure 7-2   Wiring diagram of bridge box**

(1) The measurement of tension (compression) force

As shown in Figure 7-3a, the specimen is subjected to a force $P$ and its direction is known. To measure the force, a working resistance strain gauge $R_1$ can be attached along with the force acting direction, and a temperature compensation gauge $R_2$ can be attached to another metal block of the same material in the same temperature environment with the specimen without force. Strain gauges $R_1$ and $R_2$ are connected to the bridge for measuring force $P$, as shown in Figure 7-3c. Therefore, the bridge can be mutually compensated and the output voltage is

$$U_o = \frac{1}{4}\frac{\Delta R}{R}U_i = \frac{1}{4}K \cdot \varepsilon \cdot U_i \qquad (7\text{-}1)$$

The temperature complement gauge $R_2$ can also be attached to the same test piece ( as shown in Figure 7-3b ) to form a half-bridge, as shown in Figure 7-3c. Its output voltage is increased by $1+\mu$ ( $\mu$ is Poisson's ratio) times.

$$U_o = \frac{1}{4}U_i \cdot K \cdot \varepsilon \cdot (1+\mu) \qquad (7\text{-}2)$$

Obviously, the above two kinds of patch and bridge connection methods can not exclude the influence of bending. If there is bending, it will also cause resistance changes and produce voltage output. The magnitude of the pulling force $P$ can be calculated as follows

$$P = \sigma \cdot A = E \cdot \varepsilon \cdot A \qquad (7\text{-}3)$$

Where $E$ is the elastic modulus of the specimen material; $\varepsilon$ is the measured strain value ( i.e. mechanical strain) ; $A$ is the cross-sectional area of the specimen.

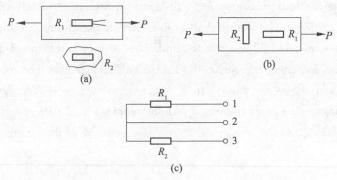

Figure 7-3   Measurement of tension (compression) load

(2) The measurement of bending load

A bending moment $M$ is applied to the test specimen, as shown in Figure 7-4a. The working resistance strain gauge $R_1$ is attached to the test specimen. The temperature compensation gauge $R_2$ is attached to a stress-free material with the same environmental temperature as the specimen. Connect $R_1$ and $R_2$ (as shown in Figure 7-4c) to the half-bridge, which is the measuring bridge of bending moment $M$, and its output voltage is

$$U_o = \frac{1}{4} \frac{\Delta R}{R} U_i = \frac{1}{4} U_i \cdot K \cdot \varepsilon$$

The method shown in Figure 7-4b can also be used to attach working gauges $R_1$ and $R_2$ to the test piece, which also compensates each other for temperature. $R_1$ is attached to the compression zone, and $R_2$ is attached to the tension zone. Their resistance changes are equal in magnitude and opposite in sign, forming a half-bridge as shown in Figure 7-4c. At this time, the output is twice that of the former, i.e

$$U_o = \frac{1}{2} U_i \cdot K \cdot \varepsilon$$

The bending moment $M$ can be calculated by the following formula

$$M = W \cdot \sigma = W \cdot E \cdot \varepsilon \tag{7-4}$$

Where $W$ is the bending section coefficient of the specimen.

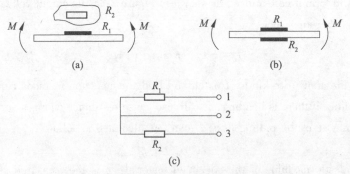

Figure 7-4   Measurement of bending moment

（3）The measurement under the combined action of tension, compression and bending

If only the bending moment is required to be measured, strain gauges and connecting bridges can be attached as shown in Figures 7-5a and b. At this time, $R_K$ is not used. Because the strain caused by tension or compression makes $\Delta R_1$ and $\Delta R_2$ equal in size and same in sign, which cancels each other on the bridge arm and will not affect the output of the bridge. Therefore, the output of the measuring bridge automatically eliminates the influence of tension (compression), which just reflects the magnitude of the bending moment $M$, and its output voltage formula is as follows.

$$U_\mathrm{o}=\frac{1}{2}U_\mathrm{i}\cdot K\cdot\varepsilon$$

If only pulling force (pressing force) is measured without considering the effect of bending, strain and connection bridges can be attached as shown in Figure 7-5a and c, $R_1$ and $R_2$ are connected in series to form an arm bridge, and the other arm is connected in series with two temperature compensation plates $R_K$. $R_K$ is attached to the unstressed part in the same environment and material as the test piece, and the output of the bridge can only reflect the magnitude of the pulling (pressing) load. The resistance changes of $R_1$ and $R_2$ caused by bending moment $M$ are equal in absolute value and opposite in sign, and cancel each other on one bridge arm, so the output of the bridge only represents pulling (pressing) load, and its output voltage formula is as follows.

$$U_\mathrm{o}=\frac{1}{4}U_\mathrm{i}\cdot K\cdot\varepsilon$$

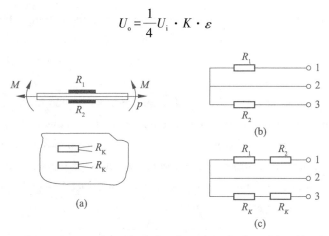

Figure 7-5   Measurement of tension (compression) and bending load

（4）The measurement of shear force

Resistance strain gauges can only measure normal stress, but not shear stress, because shear stress cannot deform resistance strain gauges and produce resistance changes. Therefore, only the normal stress caused by shear stress can be used to measure the shear force.

As shown in Figure 7-6a, $Q$ is the measured shear force, and resistance strain gauges $R_1$ and $R_2$ are pasted at $a_1$ and $a_2$, and the bending moments at these two points are respectively $M_1=Qa_1$ and $M_2=Qa_2$. According to the material mechanics, the following conclusions are consulted, $M_1 = E\cdot\varepsilon_1\cdot W$, $M_2=E\cdot\varepsilon_2\cdot W$ ($\varepsilon_1$ and $\varepsilon_2$ are strain values at $a_1$ and $a_2$ respectively). Finally, the expressions of $Q = E \cdot \varepsilon_1 \cdot W/a_1$ or $Q = E \cdot \varepsilon_2 \cdot W/a_2$ are obtained. Therefore, as long as the strain gauge is used to measure the strain value on a certain section, the transverse shear force $Q$ can be calculated.

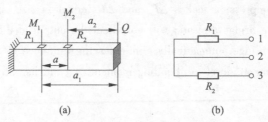

(a)                    (b)

**Figure 7-6    Measurement of transverse shear force**

The disadvantage of this scheme is that when the action point of force $Q$ changes ($a_1$ or $a_2$ changes), the measurement results will be affected. Moreover, in some cases, the value of $a_1$ or $a_2$ can not be accurately measured, but the patch position between the two strain gauges $R_1$ and $R_2$ can be accurately measured. Therefore, the above method can be changed as follows.

$$M_1 - M_2 = Qa_1 - Qa_2 = Q(a_1 - a_2) = Qa \qquad (7\text{-}5)$$

Where the coefficient $a$ is the distance between two strain gauges $R_1$ and $R_2$. It can be concluded that

$$Q = \frac{\varepsilon_1 EW - \varepsilon_2 EW}{a} = \frac{\varepsilon_1 - \varepsilon_2}{a} EW \qquad (7\text{-}6)$$

If $R_1$ and $R_2$ form a half-bridge as shown in Figure 7-6b, the output of the measuring bridge is proportional to $\varepsilon_1 - \varepsilon_2$ and has nothing to do with the change of the action point of shear force $Q$. Coefficients $a$, $E$ and $W$ are all constants, so the shear force $Q$ can be calculated by Formula (7-6).

(5) The measurement of shear stress and torque in cross-section of the shaft during torsion

According to material mechanics, when a circular shaft is subjected to a pure torque, the direction at 45° with the axis is the main stress direction, as shown in Figure 7-7a, and the absolute values of the principal stress in the vertical direction of pulling and pressing are equal and the signs are opposite, and their absolute values are numerically equal to the maximum shear stress $\tau_{\max}$ on the circumferential cross-section.

$$\sigma_1 = -\sigma_3, \quad |\sigma_1| = \tau_{\max}$$

The strain $\varepsilon$ can be measured by sticking the strain gauge on the surface of the round shaft at 45° to the axis. According to the generalized Hooke's law, $\sigma = E\varepsilon/(1+\mu)$, the maximum shear stress on the section where the strain gauge is attached is $\tau_{\max} = |\sigma| = |E\varepsilon/(1+\mu)|$. Torque is solved as follows.

$$M_K = \tau_{\max} W_P = \left| \frac{E\varepsilon}{1+\mu} \right| W_P \qquad (7\text{-}7)$$

Where $W_P$ is the torsional section modulus of the circular shaft.

In practice, two or four strain gauges are often attached perpendicularly to each other to form a half-bridge or full-bridge measuring circuit, which can increase the final output and solve the problem of temperature compensation. In engineering application, the shaft often bears bending moment as well as torque.

If the bending moment has a large gradient along the axial direction, the strain gauge sticking scheme in Figure 7-7a should be adopted instead of Figure 7-7b and Figure 7-7c. As shown in Figure 7-7d, four strain gauges are attached to the shaft. Connect $R_1$ and $R_4$, $R_2$ and $R_3$ in series, then connect to the circuit bridge during measurement, and the torque on the shaft can be tested to eliminate the influence of bending moment. If the bending moment of the bearing shaft varies greatly along the axial direction, it should be pasted and connected to the bridge according to the scheme in Figure 7-7e, in which $R_3$ and $R_4$ indicate the positions corresponding to $R_1$ and $R_2$ on the back.

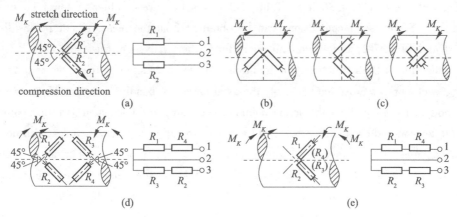

Figure 7-7    Layout of strain gauge in torque test

## 7.1.2    Stress monitoring of large structures

In the process of large equipment hoisting, the use of electrical measuring method to monitor the stress plays an important role in the safe and reliable hoisting work. Especially for some high and heavy equipment, such as chemical plant distillation tower, television tower, etc, if they are lifted in sections, there will be a lot of aerial work, which is unsafe and difficult to guarantee the quality.

For example, the propylene distillation tower in China's 300 000 t of ethylene project is 83 m high and weighs more than 500 t. Another example is that the torch tower in a project, it is 120 m high and weighs more than 230 t. These types of equipment are difficult to be hoisted in sections. Instead, they are erected by masts after they are first assembled horizontally on site. Due to the height and weight of these devices, the stress caused by the dead weight will be large when they are hoisted horizontally. Especially near the lifting point, due to the action of concentrated force, its stress is greater.

The safety of hoisting will depend on the stress of these parts. Improper operation may cause equipment damage or personal accident. If the stress of dangerous parts is monitored by electrical measurement during hoisting, accidents can be reduced or even avoided. This is an important means to ensure safety in field hoisting.

Taking the hoisting of torch tower in a 300 000 t ethylene project of a chemical company as an example, it illustrates how to carry out stress monitoring in the hoisting process. With a total height of 121.5 m and a weight of 235.6 t, the torch tower is a truss structure and is erected by the A-type mast pulling down method (see Figure 7-8).

Strain gauges are respectively arranged at each rod and bottom fulcrum near the two lifting points ($P_1$ and $P_2$) of the tower, as shown in Figure 7-9. The tower is supported with multiple sleepers before hoisting, and its levelness is measured with a level meter to minimize its stress in the initial state.

Table 7-1 shows the stress values of main stressed bars near the upper lifting point (i.e. the lifting point at the height of 86 m); Table 7-2 shows the stress values of tower columns and inclined rods. Some useful conclusions can be obtained from the measurement results listed in Table 7-1 and Table 7-2, which are convenient for further analysis of truss structure.

For this kind of truss structure, the traditional method is to simplify the joints as hinges, that is to say, each bar is a two-force bar, only the axial force, no bending moment. The measurement results show that the bars near the action points of concentrated force (such as lifting points) are not simple tension and compression bars, but also have bending stress, sometimes even dominant.

For example, the average axial stress at points 2 and 3 of the upper tower column is 16 954 MPa, while the bending stress reaches 77 714 MPa; the bending stress of inclined bar and cross bar near the lifting point also accounts for a large proportion; the average axial stress of the tower column under the outrigger is − 10 486 MPa, while the bending stress reaches ±62 524 MPa.

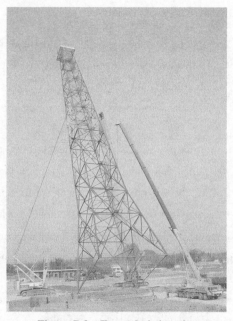

**Figure 7-8　Tower hoisting site**

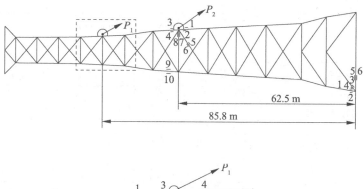

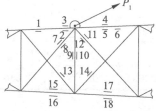

**Figure 7-9    Layout of strain gauges for torch tower**

In the area far from the lifting point, the proportion of bending stress decreases greatly. At points 15 and 16 of the lower tower column above the lifting point, the axial stress is -57 036 MPa, while the bending stress is only 392 MPa, which can be almost ignored.

From this, an important conclusion is obtained. The condition that truss joints are simplified as hinges is that each joint must be in a region without concentrated force or moment. For member bars that do not satisfy this condition, the bending stress must be considered.

Because of the concentrated force area, it is difficult to determine the theoretical simplification conditions. Therefore, in addition to the necessarily simplified calculation in advance, the field measurement of hoisting is an important and indispensable means. Through actual measurement, problems can be found at any time and major accidents can be avoided.

**Table 7-1    Stress of main load-bearing rod near the upper lifting point**

| $\varepsilon_1\sigma$ | Upper tower column(point) | | | | | |
|---|---|---|---|---|---|---|
| | 1 | 2 | 3 | 4 | 5 | 6 |
| $\varepsilon(\times10^{-6})$ | -50 | -295 | 460 | -116 | -567 | -15 |
| $\sigma$/MPa | -10 290 | -60 760 | 94 668 | -23 912 | -116 620 | -3 038 |
| $\sigma_{average}$/MPa | 16 954 | | | -70 266 | | |
| $\sigma_{bend}$/MPa | | -77 714 | +77 714 | 46 354 | -46 354 | |
| $\varepsilon_1\sigma$ | Lower tower column(point) | | | |
| | 15 | 16 | 17 | 18 |
| $\varepsilon(\times10^{-6})$ | -279 | -275 | -170 | -375 |
| $\sigma$/MPa | -57 428 | -56 644 | -34 996 | -77 126 |

continued

| $\varepsilon_1\sigma$ | Lower tower column (point) | | | |
|---|---|---|---|---|
| | 15 | 16 | 17 | 18 |
| $\sigma_{average}$/MPa | −57 036 | | −56 056 | |
| $\sigma_{bend}$/MPa | −4 | +4 | 215 | −215 |

| $\varepsilon_1\sigma$ | Inclined rod (point) | | | | | Horizontal rod | |
|---|---|---|---|---|---|---|---|
| | 7 | 8 | 11 | 12 | 19 | 9 | 10 |
| $\varepsilon(\times10^{-6})$ | −40 | −180 | −272 | −195 | 45 | −32 | 75 |
| $\sigma$/MPa | −8 232 | −37 044 | −56 056 | −39 984 | 9 310 | −6 566 | 15 386 |
| $\sigma_{average}$/MPa | −22 638 | | −48 020 | | | 4 410 | |
| $\sigma_{bend}$/MPa | 14 406 | −14 406 | −8 036 | 8 036 | | −10 976 | 10 976 |

**Table 7-2  Stress of main force-bearing rod at the leg part**

| $\varepsilon_1\sigma$ | Upper tower column (point) | | Inclined rod (point) | |
|---|---|---|---|---|
| | 1 | 2 | 3 | 4 |
| $\varepsilon(\times10^{-6})$ | −355 | 252 | −888 | 177 |
| $\sigma$/MPa | −73 010 | 52 038 | −183 064 | 36 456 |
| $\sigma_{average}$/MPa | −10 486 | | −73 304 | |
| $\sigma_{bend}$/MPa | −62 524 | 62 524 | −109 760 | 109 760 |

The following special problems must be considered and solved in the process of pressure monitoring of large structures.

### 7.1.2.1  Layout of wires

The measured areas of large structures are generally far away from measuring instruments, and the length of lead wire is usually more than 100 m. If the multi-point measurement is carried out, each strain gauge is led to the measuring instrument with long wires, which will not only cost a lot of wires but also make it difficult to find fault points once problems occur.

At this time, it is best to use the automatic patrol testing strain gauge, and place its point changing box near the measured point, so that the distance between the strain gauge and the automatic point changing box is greatly shortened, and the wire between them is controlled below 10 m. This not only saves a lot of wires but also makes it easy to find faults. Besides, wireless transmission can also be adopted, but the cost of instruments and equipment is higher.

### 7.1.2.2  Moisture resistance of strain gauges

Due to the heavy workload of field measurement, it usually takes several days for formal measurement after the strain gauge is pasted, and the problems of outdoor moisture-proof and loss prevention are very prominent. The practice has proved that normal temperature curing epoxy resin is ideal for moisture-proof. It can not only prevent moisture but also achieve the purpose of preventing damage. After the strain gauge is pasted, use a hot air blower to drive out the moisture, and then apply a moisture-proof material.

### 7.1.2.3 Temperature compensation of strain gauges

Due to poor field conditions and long measurement time, special attention should be paid to the arrangement of temperature compensation plates. Generally, partition compensation is adopted, and the area should not be too large. The temperature compensation sheet must be placed near the measured area to reduce the drift caused by temperature and reduce the measurement time. The direction of the wire of the temperature complement gauge should be consistent with that of the working plate to ensure the same temperature field of the wire.

In a word, the field measurement is not as ideal as the experiment under laboratory conditions, and there will be various problems. Problems encountered should be analyzed concretely and solved step by step.

## 7.2 Mechanical vibration test and its application

Under the excitation of some external conditions, the mechanical equipment will make a small reciprocating motion near its equilibrium position. This reciprocating motion at regular intervals is called mechanical vibration. Mechanical vibration is widespread. In most cases, mechanical vibration is often a negative or even harmful phenomenon that accompanies normal movement. It will affect the normal function and performance of mechanical equipment, such as reducing the machining accuracy of machine tools, causing accelerated wear of machine components, and even leading to rapid fracture and damage, causing accidents, and so on.

At the same time, mechanical vibration also causes mechanical equipment to emit noise, which will pollute the environment and endanger human health. However, in some cases, vibration can also be used, such as vibrating screen, vibration transmission, mechanical hammer, vibration stirrer and other mechanical equipment. On the one hand, with the development of science and technology, the requirements for the movement speed and bearing capacity of mechanical equipment are increased, which leads to the increased possibility of mechanical vibration. On the other hand, the requirements for working accuracy and stability of mechanical equipment are getting higher and higher, so the requirements for mechanical vibration control are becoming more and more urgent.

The purpose of the mechanical vibration test is to find out the vibration source or vibration transmission way through analysis, so as to minimize or eliminate the influence of vibration on the function and performance of mechanical equipment. In engineering vibration theory, the calculation method of theoretical analysis is often used to solve engineering vibration problems. The physical characteristics of the vibration system are described by using physical quantities such as mass, damping and stiffness, thus forming the mechanical model of the system.

However, in the vibration research of actual engineering structures, the structure should be simplified to some ideal mechanical models, and then analyzed. Through mathematical analysis, the modal characteristics (natural frequency, modal mass, modal damping, modal stiffness and modal vector, etc.) under free vibration are obtained. At the same time, the mechanical vibration test also includes the test of motion parameters (velocity, acceleration, amplitude, etc.).

## 7.2.1　Vibration type and its characterization parameter

### 7.2.1.1　Vibration type

Mechanical vibration refers to the process that the displacement of an observation point on mechanical equipment changes with time around its mean value or relative reference under the running state.

Similar to the classification of signals, mechanical vibration can be divided into two categories according to the vibration law: steady-state vibration and random vibration.

### 7.2.1.2　Basic parameters of vibration

Vibration amplitude, frequency and phase are three basic parameters describing the form and degree of mechanical vibration, which are called three elements of vibration.

(1) Amplitude

Vibration amplitude reflects the strength of mechanical vibration. The main forms of vibration amplitude include peak value, effective value and average value.

(2) Frequency

The frequency of vibration is generally expressed by the frequency of vibration per second $f$ (Hz) or angular frequency $\omega$ (rad/s). Simple harmonic vibration is the simplest form of periodic vibration, which has only one frequency component. Complex periodic vibration is composed of many frequency components. The main frequency components and their amplitudes can be determined by frequency spectrum analysis, which provides a basis for finding vibration sources and making vibration reduction and elimination measures.

(3) Phase

In some cases, the phase information of the vibration signal is of great significance, such as determining resonance point by using phase relation, the dynamic balance test of rotating parts, etc. The phase is generally expressed by phase angle, the unit is radian (rad) or angle (°).

## 7.2.2　Forced vibration of freedom system with single degree

The single degree freedom system is one of the simplest mechanical models. The whole mass $m$ (kg) of the system is concentrated at one point, which is supported by a spring with stiffness $k$ (N/m) and a damper with a viscous resistance coefficient $c$. It is assumed that the coefficients $m$, $k$ and $c$ do not change with time and the system is linear. The system can be expressed by the second-order differential equation with constant coefficients. Vibration research of a single-degree-of-freedom system is the foundation of a multi-degree-of-freedom system, and some practical engineering structures can be simplified as a single-degree-of-freedom system. The following describes the types of mechanical vibration with a single degree of freedom vibration system model.

### 7.2.2.1   Forced vibration caused by the force of the mass block

The typical single-degree-of-freedom system is shown in Figure 7-10.

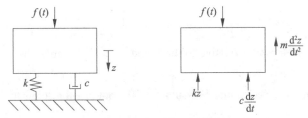

**Figure 7-10   Forced vibration of a single-degree-of-freedom system caused by force on mass block**

The motion equation of its mass block under external force is as follows

$$m\frac{\mathrm{d}^2z}{\mathrm{d}t^2}+c\frac{\mathrm{d}z}{\mathrm{d}t}+kz=f(t) \tag{7-8}$$

where, $c$ is the viscous damping coefficient; $k$ is the spring stiffness; $f(t)$ is the excitation force which is the input of the system; $z$ is the vibration displacement which is the output of the system.

Its frequency response $H(\omega)$, amplitude-frequency characteristics $A(\omega)$ and phase-frequency characteristics $\varphi(\omega)$ are as follows.

$$H(\omega)=\frac{\dfrac{1}{k}}{\left[1-\left(\dfrac{\omega}{\omega_n}\right)^2\right]+2\zeta\mathrm{j}\left(\dfrac{\omega}{\omega_n}\right)}$$

$$A(\omega)=\frac{\dfrac{1}{k}}{\sqrt{\left[1-\left(\dfrac{\omega}{\omega_n}\right)^2\right]^2+\left(2\zeta\dfrac{\omega}{\omega_n}\right)^2}} \tag{7-9}$$

$$\varphi(\omega)=-\arctan\left[\frac{2\zeta\omega/\omega_n}{1-(\omega/\omega_n)^2}\right]$$

The amplitude-frequency and phase-frequency curves can be found in the relevant content of the book *Characteristics of the Test System*.

### 7.2.2.2   Forced vibration caused by foundation motion

In many cases, the forced vibration of the vibration system is caused by foundation motion. Set the absolute displacement of the foundation is $z_1$ and the absolute displacement of mass is $z_0$. Do force analysis on them, as shown in Figure 7-11.

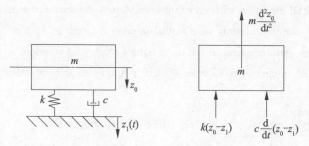

**Figure 7-11  Basic excitation of a single-degree-of-freedom system**

$$m \frac{d^2 z_0}{dt^2} + c \frac{d}{dt}(z_0 - z_1) + k(z_0 - z_1) = 0 \tag{7-10}$$

When the mass block moves relative to the foundation, its relative displacement is as follows.

$$z_{01} = z_0 - z_1 \tag{7-11}$$

Substituting Formula (7-11) into Formula (7-10) obtains the following equation.

$$m \frac{d^2 z_{01}}{dt^2} + c \frac{dz_{01}}{dt} + k z_{01} = -m \frac{d^2 z_1}{dt^2} \tag{7-12}$$

Its frequency response $H(\omega)$, amplitude-frequency characteristics $A(\omega)$ and phase-frequency characteristics $\varphi(\omega)$ are as follows.

$$H(\omega) = \frac{\left(\dfrac{\omega}{\omega_n}\right)^2}{1 - \left(\dfrac{\omega}{\omega_n}\right)^2 + 2\zeta \mathrm{j}\left(\dfrac{\omega}{\omega_n}\right)}$$

$$A(\omega) = \frac{\left(\dfrac{\omega}{\omega_n}\right)^2}{\sqrt{\left[1 - \left(\dfrac{\omega}{\omega_n}\right)^2\right]^2 + \left(2\zeta \dfrac{\omega}{\omega_n}\right)^2}} \tag{7-13}$$

$$\phi(\omega) = -\arctan\left[\frac{2\zeta(\omega/\omega_n)}{1 - (\omega/\omega_n)^2}\right]$$

Where $\zeta$ is the damping ratio of the vibration system, $\zeta = c/(2\sqrt{km})$; $\omega_n$ is the natural frequency of vibration system, $\omega_n = \sqrt{k/m}$.

The amplitude-frequency characteristic and phase-frequency characteristic curves are shown in Figure 7-12.

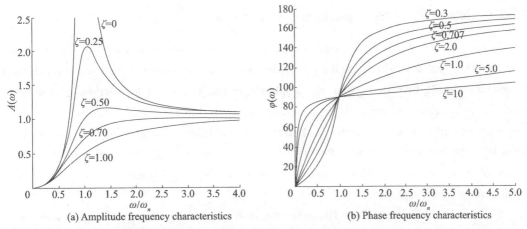

(a) Amplitude frequency characteristics       (b) Phase frequency characteristics

**Figure 7-12   Frequency response characteristics of the response to the relative displacement of the mass block to the foundation during foundation excitation**

It can be seen from the figure that when the excitation frequency is much smaller than the natural frequency of the system ($\omega \leqslant \omega_n$), the relative fundamental vibration amplitude is zero, which means that the mass block almost vibrates with the foundation, and the relative motion of the two is extremely small. When the excitation frequency is much higher than the natural frequency ($\omega \geqslant \omega_n$), $A(\omega)$ is close to 1. It shows that the relative motion (output) between the mass and the shell is almost equal to the vibration (input) of the foundation. It shows that the mass block is almost at rest in the inertial coordinates. This phenomenon is widely used in vibration measuring instruments.

It can also be seen from the figure that the response characteristics of the system are similar to those of a "high-pass" filter in terms of both high-frequency and low-frequency regions. However, in the frequency region near the resonance frequency, it is fundamentally different from the "high-pass" filter, and the output displacement is very sensitive to the change of frequency and damping.

## 7.2.3   Select the appropriate vibration sensor

A vibration sensor is a device that measures mechanical vibration parameters and converts them into electrical signals. The vibration sensor should not only have high sensitivity, flat amplitude characteristics in the measured frequency range, and the phase in a linear relationship with the frequency, but also have a small mass and small volume.

According to the different vibration parameters and frequency response ranges, vibration sensors are often divided into three types: vibration acceleration sensors, vibration velocity sensors, and vibration displacement sensors. Typical frequency response ranges are $0 \sim 50$ kHz for vibration acceleration sensors, $0 \sim 10$ kHz for vibration speed sensors and $10 \sim 2$ kHz for vibration displacement sensors. The following describes the selection parameters of vibration sensors.

### 7.2.3.1 Selection of direct measurement parameters

In the process of vibration signal measurement, the measured indicator is displacement, velocity or acceleration, which are the geometric series of $\omega$, and can be converted between them by calculus circuit. Considering that the amplitude of acceleration at low frequency may be as small as the measurement noise, if the accelerometer is used to measure the displacement of low-frequency vibration, it will make the measurement unstable and increase the measurement error because of the low signal-to-noise ratio, so it is more reasonable to use the displacement vibration sensor directly. A similar situation occurs when measuring high-frequency displacement with displacement vibration sensors.

Sensors should also be selected to make the most important parameters can be measured in the most direct and reasonable way. For example, when investigating the damage or failure caused by inertial force, it is advisable to measure acceleration; when investigating the vibration environment, it is advisable to measure the vibration speed; when monitoring the position change of machine parts, eddy current or capacitance sensor should be selected for displacement measurement. We also need to pay attention to the feasibility of sensor installation in actual machines and equipment.

### 7.2.3.2 Performance index of the sensor

All kinds of vibration sensors are limited by their structure and have their own scope of application, the selection needs to be based on the vibration frequency range of the measured system. Generally, the inertial vibration sensor with mass weight has low upper limit frequency and high sensitivity, and the inertia vibration sensor with light weight has high upper limit frequency and low sensitivity.

For the calculus amplifier, the measurable frequency range of acceleration, velocity and displacement decreases with the increase of integration times, which should be paid full attention to in use. Therefore, when choosing sensors and integrating circuits, it is necessary to consider whether to choose the analog integrating circuit or digital integrating circuit. Digital integration is to use a computer or chip to integrate the measured signal by digital integration algorithm, which has the advantages of convenience, flexibility, integration on-demand, etc. Its real-time shortcomings have been greatly reduced with the development of microelectronics.

### 7.2.3.3 Relevant precautions for use

Although laser vibration measurement has high resolution and measurement accuracy, it is only suitable for laboratory precision measurement or calibration because of extremely strict requirements on the environment and extremely expensive equipment. Eddy current sensors and capacitance sensors are non-contact, the former has low environmental requirements and is widely used in the measurement of machine vibration in industrial field. For example, the vibration sensor used in vibration monitoring of large turbo-generator sets and compressor sets should be able to work reliably in the environment of high temperature, oil pollution and steam medium for a long time, and Eddy current sensors are often used.

For vibration test items with strict requirements on phase (such as measuring virtual and real frequency spectrum, amplitude and phase diagram, mode shape, etc.), attention should be paid not only to the phase-frequency characteristics of the vibration sensor but also to the amplifier. Especially the phase-frequency characteristics of network amplifier with calculus and the phase-frequency characteristics of the other instruments in the test system, because the phase difference between the measured excitation and response includes the phase shift of all instruments in the test system.

## 7.2.4    Vibration test system

Mechanical vibration test systems are usually composed of excitation system, measurement system and analytical system, as shown in Figure 7-13. The excitation system is used to stimulate the mechanical structure under test to produce vibrations. The device used in the vibration system is called the excitation device. The measuring system includes sensors and intermediate conditioners, and the function is to convert, amplify, display or record measurement results. The analysis system can process the measured results and obtain various parameters or charts according to the research purpose.

**Figure 7-13    Composition of the mechanical vibration testing system**

### 7.2.4.1    Vibration exciting method

In structural dynamic characteristics testing, the first step is to excite the object under test, so that it can make forced vibration or free vibration according to the test requirements, so as to obtain the corresponding excitation and response signal. In engineering practice, the most commonly used excitation modes are steady state sinusoidal excitation, random excitation and transient excitation.

( 1 ) Steady-state sine excitation

Steady-state sinusoidal excitation, also known as simple harmonic excitation, is a kind of vibration excitation mode in which a single stable frequency (controllable frequency) sinusoidal excitation force is applied to the measured object through vibration excitation equipment. In engineering commonly used scanning sinusoidal excitation—sweep frequency excitation, and the excitation frequency changes with time.

( 2 ) Random excitation

Random excitation is a method of wideband excitation, it generally uses white noise or pseudo-random signal generator as a signal source. When the white noise signal passes through the power amplifier and controls the excitation equipment to generate a wideband random excitation force, the wideband random vibration response of the measured object can be aroused in the selected frequency range.

(3) Transient excitation

Transient excitation is a transient force applied to the measured object, which belongs to the wideband excitation method just like random excitation. The common transient excitation modes are fast sinusoidal scanning excitation and pulse excitation. The excitation signal of fast sinusoidal scan excitation is supplied by a signal generator whose oscillation frequency can be controlled. Usually, linear sinusoidal scan excitation is used. The signal frequency of the excitation increases linearly in the scanning period $T$, but the amplitude remains constant. Pulse excitation, also known as hammering method, is a kind of excitation method that measures excitation and response simultaneously with an impact force acting on the measured object.

In the actual pulse exciting, the pulse hammer is often used to hit the object under test, and the pulse hammer is loaded with a force sensor. The acting force of the pulse hammer on the measured object is not an ideal pulse function $\delta(t)$, but an approximate half-sine wave. Its effective frequency range depends on the pulse duration $\tau$, which depends on the hammer material. The harder the hammer material is, the smaller the duration $\tau$ is, the larger the frequency range is. Therefore, the required frequency bandwidth can be obtained by using appropriate hammer material. The excitation force can be adjusted by changing the mass of the counterweight block and the percussive acceleration of the hammer head.

### 7.2.4.2  Vibrator

The vibrator is a device that applies a certain predetermined exciting force to the measured object and exciting the measured object to vibrate. Commonly used vibrator includes electric vibrator, electromagnetic vibrator and electro-hydraulic vibrator. Here is a brief introduction to their working principles.

(1) Electric vibrator

The working principle of the electric vibrator is that the charged conductor moves under the action of electromotive force in the magnetic field, which drives the measured object to make forced vibration. Figure 7-14 shows the structure of an electric vibrator. The driving coil is fixed on the ejector rod and supported in the shell by the supporting spring, and the coil is just located in the air gap between the magnetic pole and the iron core. When the coil passes the alternating current amplified by power, the coil will be affected by an electric force proportional to the current, which is transmitted to the measured object through the ejector rod. The electromotive force generated by the vibrator is generally not equal to the exciting force applied to the measured object.

The ratio of electric force to exciting force of vibrator is called force transfer ratio. The electric force of the vibrator is equal to the vector sum of the exciting force and the elastic force, damping force and inertia force of the vibrator moving parts. Only when the mass of the moving part of the vibrator is negligible compared with the object under test, and the connection stiffness of the vibrator and the object under test is good, and the stiffness of the pusher system is good, the electric force can be considered equal to the exciting force. Therefore, the push rod is usually used to excite the object through one force sensor in order to accurately measure the magnitude and phase of the excitation force.

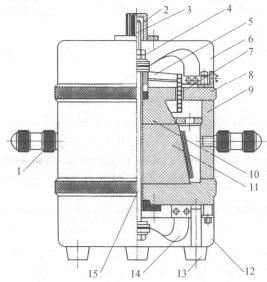

1—Credit; 2—Protective cover; 3—Connecting rod; 4—Nut; 5—Connecting skeleton;
6—Cover; 7—Coil; 8—Magnetic pole; 9—Shell; 10—Iron core; 11—Magnet;
12—Lower cover; 13—Sole foot; 14—Support spring; 15—Top bar
**Figure 7-14   Structure diagram of electric vibrator**

(2) Electromagnetic vibrator

Electromagnetic vibrator directly uses electromagnetic force as an excitation force, which is often used for non-contact excitation. The electromagnetic vibrator is composed of a core, excitation coil, force measuring coil and base, and its structure is shown in Figure 7-15. When the current passes through the excitation coil, the corresponding magnetic flux is generated, thus generating electromagnetic force between the iron core and the armature. Two main coils are sleeved on the iron core, the inner layer is DC winding, the outer layer is AC winding, and the force measuring coil is sleeved at the upper end near the magnetic gap.

The excitation force is measured by force measuring coil and the relative displacement between the vibrator and armature is measured by displacement sensor. The vibrator and armature are fixed on the measured object respectively, so that the relative excitation between them can be realized without contact. The electromagnetic vibrator has the advantages of small volume and weight and no contact with the measured object, so it can rotate and excite the measured object.

It is not affected by additional mass and stiffness, and its upper-frequency limit is about $500 \sim 800$ Hz. The disadvantage of the electromagnetic vibrator is that because the exciting force is generated in the magnetic gap, the linearity of the exciting force will be affected when the displacement amplitude is too large, and it is difficult to accurately measure the exciting force.

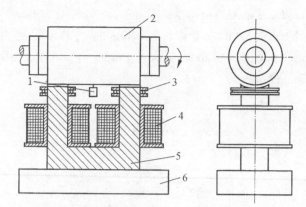

1—Displacement sensor; 2—Armature; 3—Force-measuring coils;

4—Excitation coil; 5—Iron core; 6—Base

**Figure 7-15   Structure diagram of electromagnetic vibrator**

（3）Electro-hydraulic vibrator

When exciting a large structure, in order to get a large response, sometimes a large exciting force is needed. At this time, an electro-hydraulic vibrator can be used, and its structure is shown in Figure 7-16. The signal generated by the signal generator is first amplified, and then the signal controls the action of the electro-hydraulic servo valve, making the piston do reciprocating movement, exciting the vibration of the measured object. Electro-hydraulic vibrator has large excitation force, large stroke and small volume per unit force.

However, due to the compressibility of oil and the friction of high-speed flow, the high-frequency characteristics of the vibrator are poor, which is only suitable for the lower frequency range (generally 0 ~ 100 Hz, up to 800 Hz), and its frequency characteristics are worse than those of the electric vibrator. This kind of shaker has complex structure, high manufacturing accuracy requirements, requires a hydraulic system and high cost.

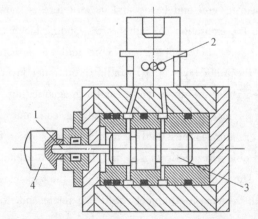

1—Ejector rod; 2—Electro-hydraulic servo valve; 3—Piston; 4—Force sensor

**Figure 7-16   Structure diagram of electro-hydraulic vibrator**

### 7.2.4.3   Vibration signal analyzer

The vibration signal detected by the vibration sensor and the excitation force signal detected by the excitation point needs to be properly processed before various useful information can be extracted. The simplest vibration analyzer indicates the vibration signal by the parameters of peak value, peak-to-peak value, average absolute value or effective value, etc. Such instruments generally include calculus circuits, amplifiers, voltage detectors, and meter heads, but they can only obtain information on vibration intensity (vibration level), but cannot obtain information on other aspects of vibration signal. In order to obtain more information, the frequency spectrum analysis is often carried out to determine the frequency components in the vibration signal, estimate the vibration source, and find the amplitude-frequency characteristic, phase-frequency characteristic or dynamic characteristic parameters of the measured system.

In vibration signal analysis, appropriate filtering and signal analysis techniques are generally used. For example, the analog spectrum analyzer consists of an amplifier, a filter and a detector. The key part is the filter, which is a frequency selection device, can make the signal of a specific frequency components pass through, inhibit or greatly attenuate other frequency components or noise. Therefore, filter is a powerful tool for frequency analysis and noise suppression. With the development of computer technology, lots of signal processing work can be completed by using signal processing software, the basic principle refers to signal processing related books.

## 7.2.5   Vibration test in mechanical condition monitoring and diagnosis

### 7.2.5.1   Diagnosis scheme

The concrete vibration test and diagnosis scheme is determined based on the comprehensive understanding for the measurement object. Whether the diagnostic scheme is correct or not is related to whether the necessary and sufficient diagnostic information can be obtained, so it must be treated carefully. A more complete field vibration testing and diagnosis scheme includes the following.

(1) Determine the measuring point

The measuring point is the part of the machine to be measured, which is the input for obtaining diagnostic information. Whether the measuring point is chosen correctly or not is related to whether the real and complete state information can be obtained. Only based on fully understanding the diagnosis object, can the measuring point be properly selected according to the diagnosis purpose. The measuring points shall meet the following requirements.

① Sensitive to vibration response

The selected measuring point should be as close to the vibration source as possible. Avoid or reduce the interface, cavity or spacer (such as sealing packing, etc.) of the signal on the transmission channel. It is best to let the signal propagate in a straight line to reduce the energy loss of the signal during transmission.

② Rich in the information

Usually, the location where vibration signals are concentrated is selected to obtain more state information.

③ Adaptation for diagnostic purposes

The selected test points should be subject to the diagnostic purpose, and the test points should also be changed if the diagnostic purpose is different.

④ Suitable for placing sensors

The measuring points must have enough space for placing sensors and ensure good contact. The measuring points should also have sufficient rigidity.

⑤ Satisfy the requirements of safe operation

Since the onsite vibration measurement is carried out while the equipment is running, the safety of personnel and equipment must be ensured when placing the sensor. For parts that are inconvenient to operate or have hidden safety hazards during operation, we must have reliable security measures; otherwise, it is best to give up temporarily.

Under normal circumstances, the bearing is the most ideal place to monitor vibration, because the vibration load on the rotor directly acts on the bearing and connects the machine with the foundation as a whole through the bearing, so the vibration signal at the bearing part also reflects the condition of the foundation.

In general, the bearing is the most ideal part to monitor vibration, because the vibration load on the rotor directly acts on the bearing, and through the bearing, the machine and the foundation are connected into a whole, so the vibration signal of the bearing part also reflects the condition of the foundation. Therefore, in the absence of special requirements, bearings are the preferred measuring point. If conditions do not permit, the measuring point should be as close to the bearing as possible to reduce the mechanical impedance between the measuring point and the bearing seat. In addition, the feet, foundations, casings, cylinders, inlet and outlet pipes, valves and other parts of the equipment are also permanent measuring points for vibration measurement. The choice must be made according to the diagnosis purpose and monitoring content.

In the field diagnosis, such a situation is often encountered, and some equipment meets great difficulties in selecting measuring points. For example, the driving mechanism of cigarette machines and packaging machines in cigarette factories is mostly enclosed in the casing, which is inconvenient to monitor the bearing parts. This situation also exists in other equipment. For example, when diagnosing a vertical drilling machine, 13 measuring points were selected, only 4 of which were close to bearings, and the others were far apart. In the case of this situation, only the other measuring parts can be selected. To completely solve the problem, some necessary modifications must be made to certain structures of the equipment according to the requirements of the inspection.

The vibration characteristics of some equipment have obvious directionality, and the vibration signals in different directions often contain different fault information. Therefore, each measuring point should generally measure three directions, namely horizontal, vertical and axial, as shown in Figure 7-17. The horizontal and vertical vibration reflect radial vibration, and the measuring direction is perpendicular to the axis. The direction of axial vibration is coincident with or parallel to the axis.

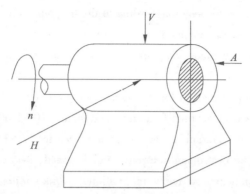

*H*—Horizontal direction; *V*—Vertical direction; *A*—Axial direction

**Figure 7-17   Three measuring directions of measuring points**

After selecting the measuring point, mark the position of the measuring point in the schematic diagram of the equipment structure. Once the points have been determined, they must often be measured at the same point. This requires permanent marking at three measuring directions at each point, such as painting or sample punching, or machining screw holes to hold the sensors. Especially for the occasion of poor environmental conditions, this point is more important. When measuring high-frequency vibration, there has been a situation where the measured value differs by 6 times after the measurement point is shifted by a few millimeters.

(2) Estimate frequency and amplitude

Before vibration measurement, a basic estimate of the frequency range and amplitude of the measured vibration signal should be made to provide a basis for selecting sensors, measuring instruments, measurement parameters, and analysis frequency bands (frequency ranges). The following simple methods can be used to estimate the vibration frequency and amplitude:

① According to the long-term accumulated field diagnosis experience, the vibration characteristic frequency and amplitude of common multiple faults are estimated.

② According to the structure characteristics, performance parameters and working principle of the equipment, calculate some possible fault characteristic frequencies.

③ The portable vibration measuring instrument is used to conduct multi-point search test in partition before formal measurement, and some parts with high vibration intensity are found, and then the sensitive frequency band and amplitude range can be roughly determined by changing the measurement frequency band and measurement parameters for multiple measurements.

④ Collect diagnosis knowledge extensively, and master fault characteristic frequency and corresponding amplitude of some commonly used equipment.

(3) Determine the measurement parameters

In vibration measurement, it is required to select the diagnostic parameters that are most sensitive to fault. This parameter is called "sensitive factor", that is, when the machine state changes in a small amount, the characteristic parameters change greatly. It is impossible to determine a sensitive factor for each fault signal because of the diversity of equipment structures and the variety of fault types. In the practice of diagnosis, a general principle is summarized, that is, the diagnosis parameters are selected according to the frequency characteristics of the vibration signal of the diagnosis object. Commonly used vibration measurement parameters including acceleration, velocity and displacement are generally selected according to the following principles:

Test the vibration displacement in low frequency occasions ($<10$ Hz).

Test the vibration speed in medium frequency occasions ($10 \sim 1\ 000$ Hz).

Test the acceleration of vibration in high frequency occasions ($>1\ 000$ Hz).

For most machines, the best diagnostic parameter is speed, because it is an ideal parameter to reflect the vibration intensity, so many international vibration diagnostic standards use the effective value of speed ($V_{rms}$) as a discriminant parameter. In the past, some industry standards in China mostly used displacement (amplitude) as a diagnostic parameter. In the selection of measurement parameters, it must be consistent with the parameters used by the discriminant standard, otherwise there will be no basis for judging the state. Besides, $g/SE$ is often used as a diagnostic parameter when diagnosing rolling bearing faults.

(4) Select the diagnostic instrument

In the selection of vibration measuring instruments, in addition to paying attention to the quality and reliability, the following two main points should be considered.

① The frequency range of the instrument should be wide enough to record all the important frequency components in the signal, generally in the range of 5 to 10 000 Hz or wider. High frequency component is an important information for predicting faults. Early mechanical fault features first appear in high frequency. When abnormalities occur in low frequency band, the fault has occurred. For the machine with very low speed, the low-frequency component is more important. It is recommended to select ultra-low frequency sensor. Therefore, the frequency range of the instrument should cover all frequency bands from high frequency to low frequency.

② The dynamic range of the instrument shall be considered. The measuring instrument is required to ensure a certain display (or recording) accuracy for all possible vibration values from the highest to the lowest within a certain frequency range. This numerical range that can ensure a certain accuracy is called the dynamic range of the instrument. For most machines, the vibration level usually varies with frequency.

(5) Selecting and installing sensors

There are three kinds of sensors used to measure vibration signals, which are generally selected according to the types of measured parameters: eddy current displacement sensors are used to measure displacement, electric speed sensors are used to measure speed, and piezoelectric acceleration sensors are used to measure acceleration.

Due to the wide frequency response range of piezoelectric acceleration sensor, it is commonly used to measure displacement, velocity and acceleration at the same time without special requirements.

Vibration measurement not only has strict requirements on the performance and quality of sensors but also pays attention to their installation forms. Different installation forms are suitable for different occasions. Table 7-3 shows the performance comparison of several common installation forms of piezoelectric accelerometer, among which the screw connection is the most ideal. However, in field measurement, especially for large-scale census test, the permanent magnetic seat is the most common method because it is the easiest to install and its performance is moderate.

**Table 7-3    Common installation methods and characteristics of piezoelectric accelerometer**

| The installation method and frequency response range ( ±3 dB) | Advantages | Disadvantages |
| --- | --- | --- |
| Handheld steel probe: 1~1 000 Hz  Handheld aluminum probe: 1~700 Hz | Attaching quickly; Suitable for all kinds of surfaces | Limited frequency range; Pay attention to the handheld method |
| Carbon block: 1~2 000 Hz | Attaching quickly | Limited frequency range; The machine must have a ferromagnetic surface, which must be clean |
| Threaded connection: 1~10 000 Hz | Wide available frequency range; The measurement reproducibility is the best | It takes time to have screw hole joints |

Before test, the sensor's performance indicators must be tested and qualified.

It should also be noted that there are two different methods for measuring rotor vibration signal, i.e. measuring absolute vibration signal and measuring relative vibration signal, as shown in Figure 7-18. The bearing vibration excited by the rotor alternating force is called absolute vibration; under the action of exciting force, the vibration of rotor relative to bearing is called relative vibration. The piezoelectric accelerometer is used to measure absolute vibration, while eddy current displacement sensor is used to measure relative vibration of rotor. The simple vibration diagnosis on-site is mainly used to measure the absolute vibration of bearings by piezoelectric accelerometer.

(6) Make preparations for other related matters

The preparation before measurement must be careful. In order to prevent measurement errors, it is best to do a simulation test before the formal measurement to verify that the instrument is in good condition and that the preparation is adequate. For example, it may seem trivial to check whether the electric power of the instrument is sufficient, but it should not be neglected. In the field, it often happens that the diagnostic work has to be stopped because of the lack of electric power of the instrument. All kinds of record forms should also be prepared so that everything is ready.

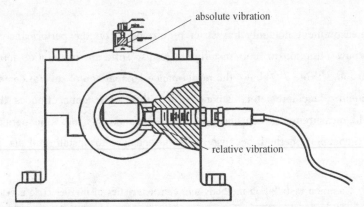

absolute vibration

relative vibration

**Figure 7-18    Schematic diagram of absolute and relative vibration of rotor**

### 7.2.5.2    Test and signal analysis

(1) Measuring system

At present, the field vibration measurement system takes two basic forms, and its structure and characteristics are described as follows.

① Measurement system composed of analog vibrometer. In the early stage of equipment diagnosis in Chinese enterprises (i.e., the 1980s), analog vibrometer was widely used for on-site simple vibration diagnosis, and its basic function was to measure the vibration parameters of machines and judge whether the equipment was faulty or not. When it is necessary to further analyze the equipment state, a simple oscilloscope and a simple frequency analyzer can be added to form a simple measurement system. It can not only observe the vibration waveform but also make a simple frequency analysis in the field. This simple measurement and analysis system can solve a lot of problems in the field diagnosis, and has played a great role, even now it still has its value.

② The vibration diagnosis and measurement system is composed of digital vibration measuring instruments represented by data collector. With the development of equipment diagnosis technology in the late 1980s and early 1990s, the portable multifunctional vibration measuring instrument represented by data collector had been widely popularized and applied in enterprises, gradually replaced the analog vibration measuring instrument, became the protagonist of field diagnosis, and revolutionized the diagnosis technology. Its simple operation method and rich functions are beyond the reach of analog instruments.

(2) Signal analysis

After determining the vibration measurement and diagnosis scheme, the relevant parameters of the equipment are measured according to the test purpose. The measured parameters must include the parameters adopted in the standard, so as to be used in state recognition. If there are no special circumstances, each measuring point must measure the vibration values in horizontal ($H$), vertical ($V$) and axial ($A$) directions. For the first measured signal, signal replay and visual analysis should be carried out to check whether the measured signal is true or not. If you know the measured signal and its characteristics clearly, you can roughly judge the authenticity of the measured signal on the spot. In case of lack of information and experience, retest and trial analysis shall be conducted for many times, and records shall be made after confirming that the test is correct.

If the instrument used has the function of signal analysis, after the measurement is completed, the signal can be further analyzed in time domain and frequency domain, especially for those measuring points with abnormal vibration. The measured data must be recorded in detail and standardized in the form.

In addition to recording the signals displayed by the instrument, other contents related to the latest measurement and analysis should also be recorded, such as ambient temperature, power parameters, instrument model, number of channels in the instrument, and operating condition parameters of the equipment during measurement (such as load, speed, inlet and outlet pressure, bearing temperature, sound, lubrication, etc.). If not recorded in time, wrong parameters will seriously affect the accuracy of analysis and judgment.

For the measured signals, it is best to sort them out, for example, according to all directions of each measuring point, so that it is easy to extract features and find the change of signals. The data measured regularly by one equipment can also be counted together, which is conducive to comparative analysis.

### 7.2.5.3   State discrimination

According to the measured signal and the feature information obtained by feature extraction, the running state of the equipment can be judged. First judge whether it is normal, then further analyze the abnormal equipment, and point out the cause, location and serious degree of the fault. For those difficult faults that cannot be solved by simple diagnosis, precise means must be used to diagnose them.

### 7.2.5.4   Diagnostic decision

Through several steps such as signal analysis and state identification, the actual state of the equipment is made clear, which creates conditions for maintenance decision-making. At this time, suggestions should be put forward for treatment: either continue to operate, or shut down for repair. For the equipment to be repaired, the specific contents of repair shall be pointed out, such as the fault parts to be handled, the parts to be replaced, etc.

### 7.2.5.5   Reexamination

The whole process of equipment diagnosis does not end with a conclusion. At the end, there is an important step to reexamine the diagnosis conclusion and the results of processing decisions. The diagnostic personnel shall learn the details of equipment disassembly and maintenance and the effect after treatment from the user. If possible, it is best to visit the site and check whether the diagnosis conclusion is consistent with the actual situation. This is the most authoritative summary of the whole diagnosis process.

# 7.3    Mechanical displacement test and its application

Displacement measurement can be divided into static displacement and dynamic displacement according to the characteristics of measurement parameters. Many dynamic parameters such as force, torque, speed, acceleration are based on displacement measurement.

When measuring displacement, appropriate measuring points, measuring directions and measuring systems should be selected according to different measuring objects. Among the components of the measurement system, the difference of sensor performance has the most prominent influence on measurement. Some commonly used displacement sensors and their performance characteristics are shown in Table 7-4. We can have a general understanding of the displacement sensors from the table.

**Table 7-4    Common displacement sensors and their characteristics**

| Type | | Measuring range | Accuracy | Performance characteristics |
|---|---|---|---|---|
| Sliding wire resistance type | Linear displacement | $1 \sim 300$ mm | ±0.1% | Simple structure, convenient use, large output and stable performance |
| | Angular displacement | $0° \sim 360°$ | ±0.1% | The resolution ratio is low and the noise of the output signal is large, so it is not suitable for dynamic measurement at high frequency |
| Resistive strain chip | Straight-line type | ±250 μm | ±2% | Solid structure, stable performance and good dynamic characteristics |
| | Swing angle type | ±12° | | |
| Inductive type | Variable air gap type | ±0.2 mm | | The structure is simple and reliable, and it is only used for small displacement measurement |
| | Differential transformer type | $0.08 \sim 300$ mm | ±3% | The resolution is good and the output is large, but the dynamic characteristics are poor |
| | Eddy current type | $0 \sim 500$ μm | ±3% | Non-contact, easy to use, high sensitivity and good dynamic characteristics |
| Capacitive | Variable area type | $10^{-3} \sim 100$ mm | ±0.005% | The structure is very simple, the dynamic characteristics are good, but it is easily affected by temperature, humidity and other factors |
| | Variable gap type | $0.01 \sim 200$ μm | ±0.1% | The resolution is good, but the linear range is small, and other characteristics are the same as the variable area type |

continued

| Type | | Measuring range | Accuracy | Performance characteristics |
|---|---|---|---|---|
| Hall element type | | ±1.5 mm | ±0.5% | Simple structure, good dynamic characteristics and poor temperature stability |
| Inductive synchronizer type | | Few meters | 2.5 μm/ 250 mm | Digital, simple structure, convenient extension, suitable for large displacement static and dynamic measurement, used for automatic detection and CNC machine tools |
| Scientific diffraction gratings | Long grating | Few meters | 3 μm /1 m | Digital, high measurement accuracy, suitable for large displacement dynamic measurement, used for automatic detection and CNC machine tools |
| | Circular grating | 0° ~ 360° | ±0.5° | |
| Angle encoder | Contact | 0° ~ 360° | $10^{-6}$ rad | Good resolution and high reliability |
| | photoelectric | 0° ~ 360° | $10^{-8}$ rad | |

Due to the different requirements for the accuracy of displacement measurement in different occasions, and the different magnitude and frequency characteristics of displacement parameters, various displacement sensors and their corresponding measurement circuits or systems are naturally formed.

## 7.3.1   Common displacement sensor

The sensor that converts the measured non-electric quantity change into the coil mutual inductance change is called mutual inductance sensor. Because this sensor is made according to the basic principle of the transformer, and its secondary windings are connected in differential form, it is also called differential transformer sensor. It has many structural forms, such as variable gap type, variable area type and solenoid type, but its working principle is basically the same.

Capacitive displacement sensor is a kind of sensor which converts displacement change into capacitance change. It has the advantages of simple structure, high resolution, non-contact measurement, and can work under harsh conditions such as high temperature, radiation and strong vibration.

The measuring principle of strain gauge displacement sensor is to use an elastic element to convert the displacement into strain variable, and then use strain gauge to measure and record. The displacement sensor composed of elastic element and strain gauge is called strain gauge displacement sensor. There are many kinds of elastic elements, and the difference lies in the structural form of elastic elements. Commonly used elastic elements include cantilever beam, circular ring and half-circular ring, etc.

## 7.3.2 Application of displacement test in mechanical engineering

### 7.3.2.1 Measurement of radial motion error of rotating shaft

The motion error of rotating shaft refers to the additional motion caused by the deviation of the rotating axis from the ideal position during the rotating process. The measurement of the error motion of the rotating shaft is very important in mechanical engineering. It is of great significance not only for the motion accuracy of precision machine tool spindle, but also for the safe operation of large and high-speed units.

The motion error is the movement parallel to the axis and perpendicular to the axis at any point on the rotation shaft. The former movement is called the end face motion error, and the latter is called the radial motion error. The following is a brief introduction to the principle and evaluation of the bidirectional measurement method for the rotary error of the spindle in the machine tool.

Figure 7-19 shows the bidirectional test system for the radial error motion of the machine tool spindle. A precision ball with adjustable eccentricity is installed at the front end of the machine tool spindle through a wobble plate as a measuring reference ball, and its surface is used to reflect the rotation axis of the spindle. Sensors $T_{x1}$, $T_{x2}$ and $T_{y1}$ are installed on the centerline passing through the reference ball and perpendicular to the rotation axis of the spindle, and measure the radial error of the spindle against the reference ball.

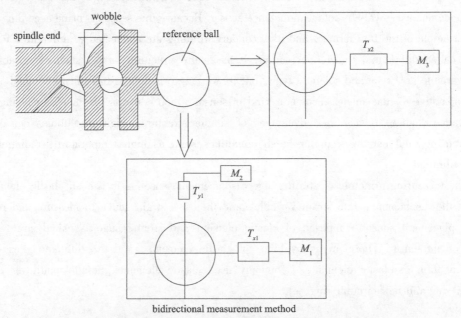

bidirectional measurement method

$T_{x1}$, $T_{x2}$, $T_{y1}$—Displacement sensors; $M_1$, $M_2$, $M_3$—Measuring instrument

**Figure 7-19   Bidirectional test system for the machine tool spindle**

In the Figure 7-20, $O_o$ is the average rotation center in the radial section of the spindle; $O_m$ is the geometric center of the reference ball; $O_r$ is the average rotation center of the spindle at a certain instant; $e$ is the installation eccentricity of the reference ball; $\delta$ is the radial motion error of the spindle at a certain instant; $\theta_0 = \omega t$, $\theta_0$ is the phase angle of a certain instantaneous motion vector, $\omega$ is the circular frequency of spindle rotation. $\theta_e$ is the angle between instantaneous error motion $\vec{\delta}$ and the eccentricity $\vec{e}$; $R$ is the distance from the fixed center point $O_0$ to the instantaneous rotation axis $O_r$. Under considering the shape errors $S_x$ and $S_y$ of the reference ball, the displacement signals detected by the displacement sensors $T_{x1}$ and $T_{y1}$ at a certain instant are respectively:

$$x = e_x + \delta_x + S_x$$
$$y = e_y + \delta_y + S_y$$

(7-14)

If the shape error of the reference ball is far less than the motion error of the spindle to be measured, for example, $s \leqslant \delta/10$, the circle image will actually be the trajectory of the sum vector $\vec{R}$ (the sum of $\vec{e}$ and $\vec{\delta}$). Then, the equation can be approximately written as follows.

$$x \approx R_x = e_x + \delta_x = e\cos(\theta_0 + \theta_e) + \delta\cos\theta_0$$
$$y \approx R_y = e_y + \delta_y = e\sin(\theta_0 + \theta_e) + \delta\sin\theta_0$$

(7-15)

At this time, the radius of gyration R can also be approximately written as follows.

$$R = \sqrt{R_x^2 + R_y^2}$$
$$= \{e^2 + \delta^2 + 2e\delta[\cos(\theta_0 + \theta_e)\cos\theta_0 + \sin(\theta_0 + \theta_e)\sin\theta_0]\}^{\frac{1}{2}}$$

(7-16)

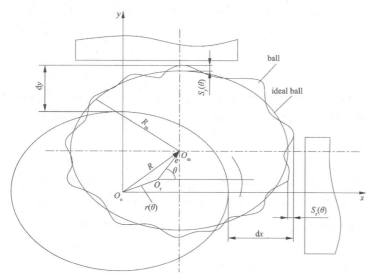

**Figure 7-20    Displacement signal analysis**

According to Equation (7-16), if $\theta_e = 0°$, when $\vec{e}$ and $\vec{\delta}$ are in the same direction, $R = e + \delta = R_{max}$, so $\delta = R_{max} - e$. If $\theta_e = 180°$, that is when $\vec{e}$ and $\vec{\delta}$ are reversed, $R = e - \delta = R_{min}$, only under the above conditions can the value of $\delta$ be determined according to the values of $e$ and $R$.

$$\Delta R = R - e = \{e^2 + \delta^2 + 2e\delta[\cos(\theta_0 + \theta_e)\cos\theta_0 + \sin(\theta_0 + \theta_e)\sin\theta_0]\}^{\frac{1}{2}} - e$$

(7-17)

It can be seen from Equation (7-17) that the circular image measured by the two sensors only reflects the variation of the radius $\Delta R$ of gyration $R$, and is not the true radial motion error $\delta$ of the spindle. It can be seen that the rotating shafts with the same value $\delta$ have different circle images because they use different eccentricity $e$, or even if the eccentricity $e$ is the same but the eccentricity orientations are different, the circle image will be different. Therefore, in general, the circular image measured by the bidirectional method cannot accurately evaluate the radial error motion of the spindle.

To sum up, the main points of measuring the radial motion error of the spindle in the machine tool with the bidirectional method can be summarized as follows.

① The gyration radius $R$ or the variation $\Delta R$ can be determined only when the value of eccentricity $e$ is known exactly and $S_x$ and $S_y$ can be ignored. Therefore, as a reference ball for measurement, its shape error must be much smaller than the motion error of the spindle, so that the influence of the spherical shape error can be ignored.

② Under normal circumstances, it can be known from Equation (7-14) that $x+y=\delta$, and only when both $S_x$ and $S_y$ tend to zero or are known, can $\delta$ be determined by $x$ and $y$. Therefore, how to eliminate or separate the eccentricity $e$ and the shape error $S$ of the reference ball has become an important task to construct the measurement method.

③ When the eccentricity $e$ of the reference ball is zero, or when $e$ is far less than $\delta$, if the signals measured by two displacement sensors with the same sensitivity are superimposed on a base circle, a circle image that truly reflects the motion track of the spindle center $O_r$ can be obtained. Also, only in this case, or under the condition of specifying a unified $e$ and a unified eccentric orientation of the reference ball, can the rotation performance of spindle be evaluated and compared according to the circular image measured by the bidirectional method.

④ Usually, proper mechanical devices and fine adjustments are used to reduce the installation eccentricity, or a filtering method is used to reduce the influence of eccentricity.

### 7.3.2.2　Level measurement

The level is the general term for liquid level, material level and interface position. The level measuring meter is used to continuously or intermittently measure the height of the fluid or solid material in the container. The purpose is to accurately measure the capacity or weight of the material stored in the containers, at any time know the height of the object level in containers, and give an alarm to the upper and lower limits of the level. Continuously monitor and adjust the production process to maintain the level at the required height.

(1) Acoustic level measurement

Acoustic level measurement is that the probe sends an ultrasonic pulse to the material surface, and when the ultrasonic pulse reaches the material surface, the material reflects it to the probe. The probe is installed in the range of the container cover. According to the ultrasonic pulse transmission time, the distance between the probe and the material surface is calculated by the microprocessor, so as to obtain the height of the material level, as shown in Figure 7-21.

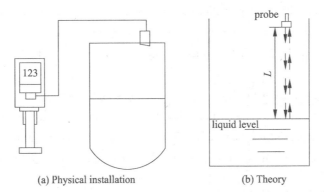

(a) Physical installation              (b) Theory

**Figure 7-21   Schematic diagram of acoustic level measurement**

Acoustic level measurement technology can be used for liquid and solid level measurement. As the probe does not contact with the measured material, the probe will not appear wear, corrosion and other phenomena. With a special probe and good measuring conditions, the maximum measuring range is up to 60 m, the maximum temperature is up to 150 ℃ and the pressure is up to 500 000 Pa.

When dust, steam and condensate are serious, the maximum measurement range will be reduced. The premise of acoustic level measurement is that the medium between the probe and the material to be measured should be air, if it is other gas, then the sound velocity should be calculated (reference measurement). If there is a heterogeneous gas layer between the probe and the material being measured, this measurement method is ineffective.

(2) Capacitive level measurement

The capacitive method can be used to measure the surface of mucous, granular and powdery materials. Capacitive level measurement technology is to measure the capacitance between the probe and the inner wall of the container, between two probes or between the probe and the concentric measuring tube. The level measurement can be realized by the deviation between the dielectric constant of liquid and solid and that of gas. However, this measurement method needs to be calibrated, and the changing dielectric constant should be compensated during continuous measurement.

In the capacitance method, a capacitance probe is used to sense the level change of the object surface. When measuring, the upper part of the capacitor is separated by air, and the lower part is filled with liquid or other materials. The dielectric constant of air is $\varepsilon_0 = 1$, and the dielectric constant of the measured object is $\varepsilon_r$. When the level of the object changes, the capacitance change value $\Delta C$ of the capacitor is linearly related to the height $x$ of the measured object.

$$\frac{\Delta C}{C_0} = \frac{x(\varepsilon_{r-1})}{h} \tag{7-18}$$

Among them, $h$ is the total height of the capacitor; $C_0$ is the initial capacitance value.

High frequency (1 MH) measurement and improved measurement circuit can significantly reduce the influence of material conductivity and the influence of conductive attachments of adhesive materials. In addition, the influence of attachment can be ameliorated within the probe range. Due to the condensation under the container cover, the formation of attachments is mainly in the range of the upper probe. The inactive part is realized by shielding the probe with a reference potential.

Capacitive level transmitter with microprocessor can simultaneously send out test signals for the measuring equipment self-inspection, including the probe rod itself. The probe rod is usually hollow, and there is a coaxial test wire inside. The test signal is directly transmitted to the top of the probe through the test wire, and then the test signal is sent to the peripheral measuring circuit. This design enables the probe to be detected in time in case of failure such as fracture or corrosion. Figure 7-22 shows the schematic diagram of a typical capacitive level measurement system.

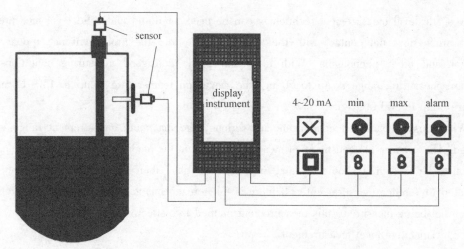

**Figure 7-22    Schematic diagram of capacitive level measurement**

Capacitive probes can be made in rod or rope type. When different materials are used, they can adapt to the harshest test conditions, especially high temperatures (400 ~ 500 ℃), high pressures, strong wear and corrosive medium.

(3) Resistance type liquid position meter

The resistance level gauge (see Figure 7-23) is composed of two rods with high resistivity. The materials and cross-sections of the two rods are the same. The two ends are tensioned and the container is insulated with the insulating sleeve. If what is measured is a conductive medium, its resistivity is very small and negligible, and the resistance of the connecting wire is omitted, then the resistance $R$ of the whole sensor is as follows:

$$R = \frac{2\rho}{A}(L'-h) = \frac{2\rho}{A}L' - \frac{2\rho}{A}h \qquad (7\text{-}19)$$

Among them, $\rho$ is the resistivity of the rod; $A$ is the cross-sectional area of the pole; $L'$ is the full length of a great bar; $h$ is the measured liquid level height.

Set: $K_1 = \frac{2\rho}{A}L'$, $K_2 = \frac{2\rho}{A}$, and then

$$R = K_1 - K_2 h \qquad (7\text{-}20)$$

In Equation (7-20), $K_1$ and $K_2$ are both constant, and the change of liquid level $h$ can be known by measuring the variable of $R$. The resistance $R$ can be measured by the circuit bridge, and the error caused by temperature change can be compensated in the circuit. However, the biggest disadvantage of this liquid level meter is that rusting, surface polarization, scaling, corrosion and other conditions on the electrode surface will change the surface contact resistance, thus directly introducing measurement error.

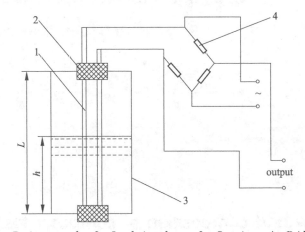

1—Resistance pole; 2—Insulation sleeve; 3—Container; 4—Bridge

**Figure 7-23   Resistance liquid position meter**

# Questions

7.1   Please illustrate the specific application of the stress-strain test with examples.

7.2   How to take measures to solve the problems of moisture resistance and temperature compensation of strain gauges?

7.3   If the working speed of a rotating machine is 3,000 rad/s, in order to analyze the dynamic characteristics of the unit, the highest frequency to be considered is 10 times of the working frequency.

(1) What type of vibration sensor should be selected, why?

(2) What is the minimum sampling frequency used for analog-to-digital conversion?

7.4   Which parts of the vibration test system are composed of, what is the role of each part?

7.5   Please classify and summarize all kinds of excitation equipment and excitation methods, and point out their respective advantages, disadvantages and their application scope.

7.6   Please briefly describe the structure, working principle and main characteristics of the electric vibrator.

# Chapter 8

## Computer Test System and Virtual Instrument

## 8.1 Overview and composition of computer test system

### 8.1.1 Overview

In the 1980s, computer technology began to be applied to instruments. With the rapid development of computer technology, large-scale integrated circuit technology and communication technology, the combination of sensor technology, communication technology and computer technology, the relationship between computer and test technology has fundamentally changed. Computer has become the basis of modern test and measurement system.

The computer test system is divided into four parts: microcomputer or microprocessor, the measuring instrument or equipment, interface and software.

Microcomputer or microprocessor is the core of the entire test system. Under the control of software, the test instrument with microprocessor is regarded as the core controls data acquisition. The multi instrument controlled by microcomputer forms the test system to calculate, transform, data process and error analysis the measured data. Finally, the measured results are stored or printed, displayed and output.

The work of the measuring instrument or system, such as the selection and adjustment of the measuring function, working frequency band, output level, range, etc., is completed under the control commands issued by the microcomputer. This kind of measuring instrument which can accept program control and change the working state of internal circuit and complete specific tasks is called programmable control of instrument, or program-controlled instrument. The instruments are connected by various buses through proper interfaces. Obviously, the interface is an important part in effective communication between the instruments and equipment of the test system, so as to realize automatic test. The main task of the interface is to provide mechanical compatibility, logic level matching, and to exchange electrical signal information through data buses.

The computer test system is divided into three parts: data acquisition and storage, data analysis and data display. In some test systems, the data analysis and display are completed by the software of microcomputer. Therefore, as long as a certain amount of data acquisition hardware is provided, the measuring instrument can be composed with microcomputer. This kind of measuring instrument based on microcomputer is called virtual instrument.

Testing technology and computer technology are almost synchronized and coordinated to develop forward. Computer technology has become the core of testing instruments and testing systems. Without the development of computer, software, network and communication, the progress of testing technology is beyond logic and above reason. At present, the computer-based test system can be divided into three types:

The first is the computer plug-in test system. That is to insert signal conditioning, analog signal acquisition, digital input and output, DSP (digital signal processing chip) and other test and analysis boards in the computer's expansion slot (usually PCI, ISA bus slot, etc.), which can also be designed into the portable computer special PCMCIA card, to form a general or special test system.

The second is the combination of instrument front end and computer. The front end of the instrument is generally composed of signal conditioning, analog signal acquisition, digital input and output, digital signal processing, test control and other modules. VXI, PXI and other special instrument buses are connected together to form an independent chassis, and are connected to the computer through communication interfaces such as Ethernet interface, 1394, and parallel interface, forming a general or special test system.

The third type is a test system composed of various independent programmable instruments (computer interfaces with parameter setting and control functions) connected to a computer. This type of system is also called an instrument control system.

The above three types of computer test systems can use general test analysis software to form a computer test system, or special software systems to form a virtual instrument. With the continuous development of microelectronics technology, VLSI chips which integrate CPU, memory, timer/counter, parallel and serial interface, encryption module on the interface and other circuits on one chip appear constantly. Taking single chip microcomputer as the main body, the computer technology and measurement control technology are combined together to form the so-called "intelligent measurement control system", that is, intelligent instruments.

## 8.1.2   Basic composition

The computer test system is shown in Figure 8-1. Compared with the traditional test system, the computer test system can convert the analog signal output by the sensor into digital signal, and achieve the purpose of test automation and intelligence by using the abundant software and hardware of the computer system.

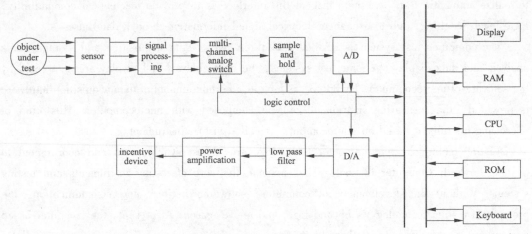

Figure 8-1   Computer test system

## 8.2   Plug-in test system

Traditional instrument is mainly composed of control panel and internal processing circuit, while the plug-in instrument does not have instrument panels. It should establish graphical virtual panel with the help of powerful graphic environment of computer to complete the instrument control, data analysis and data display. Taking the data acquisition card as an example, it usually has functions of A/D conversion, D/A conversion, digital I/O and counter/timer, and some also have the functions of digital filtering and digital signal processing. Nowadays, the multi-functional data acquisition card mostly adopts the technology of "virtual hardware". Its idea is derived from programmable devices, allowing users to easily change the functions or performance parameters of the hardware through programs, thereby relying on the flexibility of hardware devices to enhance its applicability and flexibility.

Flexible virtual instrument can be established by using general microcomputer, which is a popular virtual instrument system at present. In this way, the data acquisition card inserted into the microcomputer or industrial computer is combined with special software to complete the test task. It makes full use of the computer bus, case, power supply and system software. The key to the performance of this kind of system is the A/D conversion technology.

There are many types of plug-in cards, such as ISA card, PCMCIA card and PCI card. With the development of computer, ISA card has been gradually out of the stage. The engineering application of PCMCIA card is affected by its weak structural connection strength. The PCI bus is widely used and has become the de facto standard for microcomputers.

The PCI bus is a synchronous 32-bit or 64-bit local bus independent of CPU. Its clock frequency is 33 MHz, and its data transmission speed is as high as $132 \sim 264$ MB/s. The unlimited read-write burst mode of PCI bus technology can send a large amount of data in an instant. Peripherals on PCI bus can work with CPU simultaneously, which improves the overall performance. PCI bus also has automatic configuration function, so that all PCI compatible devices can achieve true "plug and play".

Since most of the plug-in instruments have no anti-aliasing filter and time-sharing sampling, special attention should be paid to the aliasing phenomenon and the phase difference between channels.

Because the number of personal computers is very large and the price of plug-in instruments is cheap, it is widely used, especially suitable for teaching departments and various laboratories.

Although the PCI bus-based test instrument has many advantages, there are also some disadvantages. First of all, you need to open the case when inserting the DAQ, which is inconvenient to operate, and the PCI slot on the host is limited; secondly, the test signal is directly input to the computer, and various on-site tested signals pose a great threat to the safety of the computer; thirdly, the strong electromagnetic interference inside the computer will also have a great impact on the measured signal. Therefore, the plug-in instrument system with serial interface bus mode has become the mainstream of cheap virtual instrument test system.

The bus used in this kind of test instrument system includes traditional RS232 serial bus, USB universal serial bus and IEEE1394 bus. RS232 bus is a serial bus used in the early stage of microcomputer. It is mature in technology and widely used. It is still suitable for virtual instrument or test system with low requirements. In recent years, USB bus has been widely used. All operating systems of Microsoft are compatible with USB, but USB bus is only used in simple test system. IEEE1394 is a kind of high-speed serial bus, which can transmit data at the rate of 100, 200 or 400 MB/s. At present, the transmission speed of IEEE1394 bus used in international testing instruments has reached 100 MB/s.

The hardware can be integrated into a collection box or a probe by using the serial port communication of the microcomputer. The software installed on the microcomputer can usually complete the functions of various instruments. Their biggest advantage is that they can be connected with notebook computer to facilitate field operation, and can be connected with desktop computer or industrial control computer, so it is very convenient for both desktop and portable applications. In particular, USB port and 1394 port have the characteristics of fast transmission speed, hot plug and easy online use, which have a bright future and will become the mainstream platform of virtual instrument with great development prospect and wide market in the future. Through different interface buses, automatic test system of different scales can be established. With the help of communication of different interface buses, virtual instruments, various electronic instruments with interface bus or various plug-in units can be deployed and formed into small and medium-sized automatic debugging systems.

In order to make the test instrument adapt to the configuration of the above various buses, companies all over the world, especially the American NI Company, developed a large number of software and hardware to meet the requirements. The main modular hardware, such as PXI modular instrument used for data acquisition, instrument control and machine vision, can flexibly build automatic test systems. If you want to know more about the details, you can access on its website *www.ni.com*, query the relevant board and software introduction.

# 8.3 Instrument bus

In addition to using general-purpose computer or industrial computer to develop test instruments, the special instrument bus system is also developing continuously, and has become a special platform for building high-precision and integrated instrument system. The architecture of high precision integrated system has experienced the development process of GPIB → VXI → PXI.

GPIB is a standard communication protocol between computer and instrument. The hardware specifications and software protocols of GPIB have been incorporated into international industry standards IEEE 488.1 and IEEE 488.2. It is the earliest instrument bus. At present, most instruments are equipped with GPIB interface following IEEE 488. A typical GPIB test system consists of a computer, a GPIB interface card and several GPIB instruments. Each GPIB instrument has its own address and is operated by computer. The instruments in the system can be increased, reduced or replaced, and only the control software of the computer needs to be modified accordingly. This concept has been applied to the internal design of the instrument. In terms of price, GPIB instruments cover from cheap to expensive instruments. However, the data transmission speed of GPIB is generally lower than 500 kB/s, which is not suitable for the application with high speed requirements. As a product of the early development of instruments, it has gradually withdrawn from the market.

VXI bus is a kind of high-speed computer bus, which is the extension of VME bus in instrument field. VXI bus has the characteristics of open standard, compact structure, strong data throughput capacity, accurate timing and synchronization, reusable modules, etc. VXI bus can be compatible with instruments from more manufacturers, so it has been widely used. After years of development, the construction and application of the VXI system has become more and more convenient, especially for the establishment of large and medium-scale automatic measurement systems and occasions that require high speed and accuracy. However, the construction of VXI bus requires chassis, zero slot manager and embedded controller, and its cost is relatively high. Its popularization and application are limited to some extent, and its main application is concentrated in the fields of national defense and military industry such as aviation and aerospace. At present, this type also has the trend of gradually withdrawing from the market.

PXI bus is based on Compact PCI and is extended from PXI bus with openness (proposed by NI company in 1997). The PXI bus conforms to industry standards and fully utilizes all the advantages of the bus in terms of mechanical, electrical and software characteristics. PXI structure is similar to VXI structure, but it has lower equipment cost, faster running speed and more compact volume. At present, the hardware and software based on PCI bus can be used in PXI system, so that PXI system has good compatibility. PXI also has a high degree of scalability. It has 8 expansion slots, while the desktop PCI system has only 3~4 expansion slots. PXI systems can be expanded to 256 expansion slots by using PCI-PCI bridges. The transmission rate of the PXI bus has reached 132 MB/s (up to 500 MB/s), which is the highest transmission rate that has been released.

As a standard test platform, PXI has many other advantages in addition to its absolute competitive advantage in price. First of all, with the increase of product complexity, the number of items under test increases correspondingly. Using PXI module, we can flexibly configure a comprehensive automatic test platform to test multiple functions at the same time, which can effectively save system test time and cost; secondly, PXI integrates timing and triggering, higher bandwidth and better cost performance, so it becomes the first choice of test platform; in addition, PXI provides a clear hybrid solution, that is, PXI can easily integrate hardware and software, including the previous generation VXI, GPIB and serial port devices with PXI new products, USB and Ethernet devices.

Firstly, compared with VXI, PXI chassis is smaller in size. For many large-scale integrated systems with complex functions, the modules it can provide are effective, so it can only be used in some unit test. Secondly, due to the lack of shielding box for each module in VXI system, PXI has poor electromagnetic compatibility and is not suitable for some occasions with high reliability requirements. In addition, compared with traditional instruments, PXI uses universal chips and technologies, and there is a gap between PXI and traditional instrument manufacturers in terms of sampling accuracy and other technical indicators. Therefore, it is a shortcut for the rapid development of PXI technology to learn from the experience of traditional instrument manufacturers and strengthen their cooperation.

**Example 1**: the SCADASIII produced by Belgium LMS is a front-end device for multi-channel data acquisition (as shown in Figure 8-2). This modular device can be expanded from four channels to hundreds of channels without affecting performance. Each four channels input module has a high-performance DSP chip, which can perform FFT spectrum, root mean square value and real-time octave analysis. Different sizes of LMS SCADASIII chassis can meet the needs of mobile test system. The LMS SCADAS III is a fully digital system, which can be calibrated by computer in modules, and is compatible with the LMS Test, Lab and LMS Cada-X. It has high performance signal conditioning function and supports a variety of sensors. The first expansion chassis can be placed 50 m away from the main box without affecting the measurement performance. The design of low noise cooling system can meet the requirements of sensitive acoustic test. Each mainframe includes a system controller connected to the host computer through the SCSI interface, a master/expansion chassis interface and a calibration module. The D-SCSI interface allows the main box to be placed 25 m away from the computer.

**Example 2**: the INV2308-8 wireless static strain tester produced by Beijing Dongfang Institute (as shown in Figure 8-3), each measuring point can be grouped by any bridge, and the electronic switch can switch the measuring point. It supports the resistance measurement function, and support single and multiple machine networking test. The system can be expanded infinitely. The instrument has built-in rechargeable battery with large capacity, and also supports a wide range of external DC power supply. The instrument has built-in large capacity memory, which can save tens of thousands of test data. Each measuring point bridge of the instrument is set independently, and each instrument has two compensation plates connected to the terminal. The compensation piece can be selected for each measuring point in the software, so the field test way is more flexible.

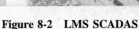

**Figure 8-2  LMS SCADAS**

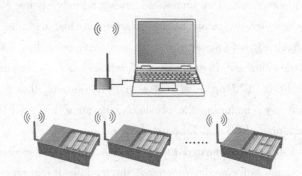

**Figure 8-3  Wireless static strain tester**

**Example 3**: the dynamic signal test and analysis system DH5922N produced by Donghua Testing Institute (as shown in Figure 8-4), which can realize multi-channel parallel sampling. Multi-channel output is not related to each other and can output a variety of signals. It can record multi-channel signals in real time and without interruption for a long time; it is equipped with a variety of programmable signal modulator, which can automatically identify each channel and normalize the data of each channel; each computer can control more than one channel synchronous parallel sampling to meet the requirements of signal acquisition with multi-channel, high-precision and high-speed.

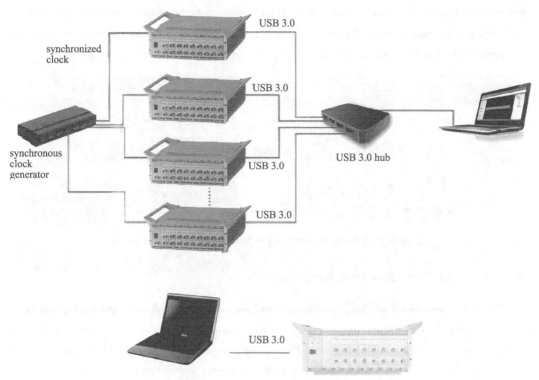

synchronized
clock

USB 3.0

USB 3.0

synchronous
clock
generator

USB 3.0

USB 3.0 hub

USB 3.0

USB 3.0

**Figure 8-4    Dynamic signal test analyzer DH5922N**

# 8.4    Introduction to virtual instrument

The main functions of measuring instruments are composed of data acquisition, data analysis and data display. In the virtual instrument system, the data analysis and display are completed by the software of microcomputer. Therefore, as long as a certain amount of data acquisition hardware is provided, the measuring instrument can be composed of microcomputer. This kind of measuring instrument based on microcomputer is called virtual instrument. In the development of virtual instrument, we can get completely different measuring instruments based on the same hardware and different software programs. It can be seen that software system is the core of virtual instrument—"software is the instrument".

Virtual instrument is an instrument that combines general technology with relevant instrument hardware through software, and is operated by users through a graphical interface (usually called a virtual front panel, as shown in Figure 8-5). The development and application of virtual instrument originated from LabVIEW software introduced by NI (National Instruments) company in 1986, and the concept of virtual instrument was put forward in that period of time. Virtual instrument makes full use of the powerful function of computer system, combined with corresponding instrument hardware, adopts module structure, which greatly breaks through the limitations of traditional instrument in signal transmission, data processing, display and storage, so that users can easily define, maintain, expand and upgrade it. It reduces the instrument cost and promotes the further combination of instrument technology and computer technology.

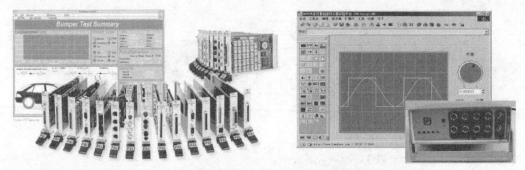

Figure 8-5　Virtual instrument hardware and instrument panel

## 8.4.1　Composition of virtual instrument

The basic components of virtual instrument include computer, software, instrument hardware and communication bus connecting computer with instrument hardware. The computer is the hardware foundation of virtual instrument. For testing and automatic control, the computer is a powerful and low-cost running platform. The virtual instrument makes full use of the graphical user interface (GUI) of the computer, and the specific application programs developed are based on the Windows operating environment, so the configuration of the computer must meet the requirements of graphical programming. GUI has the most basic requirements for the CPU speed, memory size and display card performance of the computer. Generally speaking, the computer with more than 1 GB memory can achieve good results.

In addition, the virtual instrument must be equipped with other hardware, such as various computer built-in cards or external measuring equipment and corresponding sensors to form a complete hardware system. In practical application, there are two ways to form a complete virtual instrument. One is to directly amplify and adjust the output signal of the sensor and send it to the special data acquisition card built in the PC, and then the software completes the data processing. At present, many manufacturers have developed many DAQ cards for building virtual instruments. The DAQ card can complete various functions such as A/D conversion, D/A conversion, counter/timer, etc., and with the corresponding signal conditioning circuit module, it can form various virtual instruments.

The other is to connect various test instruments with certain interfaces to the PC, such as GPIB instruments, VXI bus instruments, PC bus instruments, and instruments or instrument cards with RS-232 ports.

After the basic hardware is determined, the powerful software must be available to enable the virtual instrument to be defined as required. The software is generally composed of instrument driving software and monitoring system software. Among them, the device driver software is mainly used to complete the control program of various hardware interface functions. The virtual instrument communicates with the real instrument system through the device driver software. And it displays various controls corresponding to the real instrument panel operation elements.

In these spaces, the program-controlled information of the corresponding instrument is integrated, so the user can operate the virtual instrument panel with the mouse as real and convenient as the traditional instrument. NI company provides hundreds of drivers for GPIB, VXI, RS-232 and DAQ card. Through the instrument driver and interface software, the monitoring system software can operate the hardware, collect data, and complete functions such as data processing, data storage, report printing, trend curve, alarm and record query, drivers, as long as the user interface code and data processing software of the instrument are combined together, a new virtual instrument can be constructed quickly and conveniently. The software of the system is directly facing the operator, which requires a good man-machine interface and convenient operation. Here, the hardware completes the data acquisition and provides the specific environment for data processing, while the data processing, display and storage are completed by software. Therefore, the software is the core of virtual instrument, which defines the specific functions of the instrument.

The current popular virtual instrument software is the graphical software developing environment, and its representative products are LabVIEW and HP's VEEE. LabVIEW is aimed at general users who have no programming experience, especially suitable for engineering and technical personnel engaged in scientific research and developing. It is a graphic programming language, which simplifies the complex, tedious and time-consuming language programming into simple, intuitive and easy to learn graphic programming. The program compiled is very close to the program flow chart. Compared with traditional programming languages, using LabVIEW graphical programming can save 80% of programming time. In order to facilitate the developing, LabVIEW also provides more than 450 instrument driver libraries from more than 40 manufacturers, and integrates a large number of templates for generating graphical interfaces, including digital filtering, signal analysis, signal processing and other functional modules, which can satisfy users from process control to data processing.

In addition, HP's VEE 4.0, Lab Windows/CVI and Component Works are all excellent visual programming languages.

## 8.4.2　Application of LabVIEW virtual instrument

This section will illustrate the application and developing of virtual instruments through several examples.

**Example 8-1**　Signal generator

Figure 8-6 shows the front panel design of a dual channel signal generator. The signal generator can generate square wave, sine wave and triangle wave signal, and the frequency of the signal can be adjusted. The signal generator has the function of adding time window and spectrum analysis. Users can choose to add different time windows. The influence of window function on signal waveform and spectrum can also be observed conveniently by using this signal generator.

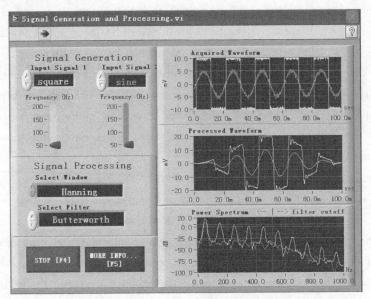

Figure 8-6　Front panel of the signal generator

**Example 8-2**　Spectrum analyzer

Figure 8-7 shows the front panel design of the spectrum analyzer. The frequency spectrum of the signal after fast Fourier transform can be displayed on an oscilloscope, and then a certain part of the frequency spectrum can be extracted and refined to obtain a more accurate spectrum.

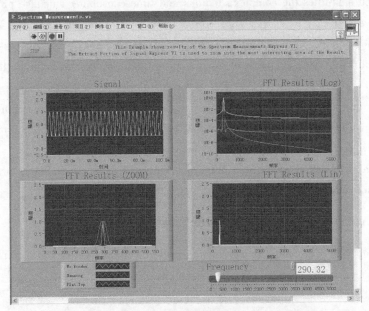

Figure 8-7　Front panel of the spectrum analyzer

**Example 8-3**　Temperature monitoring system

Figure 8-8 shows the front panel design and rear panel design of a temperature monitoring system. The system is equipped with temperature upper and lower limit alarms, when the temperature exceeds the allowable range, the system will automatically alarm and adjust. And the system can also perform statistical analysis on historical data, such as mean, standard deviation, histogram statistics.

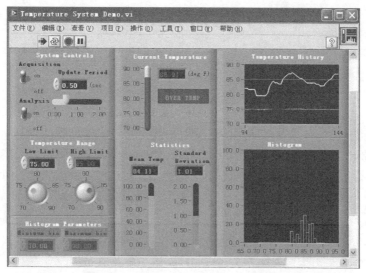

**Figure 8-8　Front panel of the temperature monitoring system**

# Questions

8.1　Briefly describe the main functions and technical requirements of each component of the computer test system.

8.2　What types can measurement systems be divided into, please explain briefly.

8.3　Briefly describe the characteristics of virtual instruments.

8.4　What aspects are included in the software developing of virtual instrument? Please explain briefly.

# References

[ 1 ] 陈保家. 机械工程测试技术及应用[M]. 北京:中国水利水电出版社,2019.

[ 2 ] 李力. 机械测试技术及其应用[M]. 武汉:华中科技大学出版社,2011.

[ 3 ] 李力. 机械信号处理及其应用[M]. 武汉:华中科技大学出版社,2007.

[ 4 ] 熊诗波. 机械工程测试技术基础[M].4 版. 北京:机械工业出版社,2018.

[ 5 ] 贾民平,张洪亭. 测试技术[M].3 版. 北京:高等教育出版社,2016.

[ 6 ] 罗志增,薛凌云,席旭刚. 测试技术与传感器[M]. 西安:西安电子科技大学出版社,
2008.

[ 7 ] 王恒. 传感器与测试技术[M]. 西安:西安电子科技大学出版社,2016.

[ 8 ] 宋雪臣,单振清. 传感器与检测技术项目式教程[M]. 北京:人民邮电出版社,2015.

[ 9 ] 陈晓军. 传感器与检测技术项目式教程[M]. 北京:电子工业出版社,2014.

[10] 吴祥. 测试技术[M]. 南京:东南大学出版社,2014.

[11] 李曼. 工程测试技术[M]. 北京:煤炭工业出版社,2017.

[12] 王三武,丁毓峰.测试技术基础[M].3 版. 北京:北京大学出版社,2020.

[13] 康宜华. 工程测试技术[M]. 北京:机械工业出版社,2005.

[14] 郑建明. 工程测试技术及应用[M]. 北京:电子工业出版社,2011.

[15] 王伯雄,王雪,陈非凡. 工程测试技术[M].2 版. 北京:清华大学出版社,2012.

[16] 吴祥. 测试技术[M]. 南京:东南大学出版社,2014.

[17] 朱先勇,于海明. 测试技术[M]. 北京:科学出版社,2019.

[18] 彭俊彬,岳建海. 测试技术[M]. 北京:北京交通大学出版社,2013.

[19] 陈光军. 测试技术[M]. 北京:机械工业出版社,2014.

[20] 刘晓彤. 测试技术[M]. 北京:科学出版社,2008.

[21] 钱苏翔. 测试技术及其工程应用[M]. 北京:清华大学出版社,2010.

[22] 祝海林. 机械工程测试技术[M].2 版. 北京:机械工业出版社,2017.

[23] 王振成,张雪松,刘爱荣,等. 工程测试技术及应用[M]. 重庆:重庆大学出版社,
2014.

[24] 陈国强,范小彬. 工程测试技术与信号处理[M]. 北京:中国电力出版社,2013.

[25] 陈花玲. 机械工程测试技术[M].2 版. 北京:机械工业出版社,2009.

[26] 李成华,栗震霄,赵朝会. 现代测试技术[M].2 版. 北京:中国农业大学出版社,2012.

[27] 杨建伟. 工程测试技术[M]. 北京:机械工业出版社,2016.

[28] 张春华,肖体兵,李迪. 工程测试技术基础[M]. 2 版. 武汉:华中科技大学出版社,
2011.

[29] 童淑敏,韩峰. 工程测试技术[M]. 北京:中国水利水电出版社,2010.

[30] 王明赞. 测试技术实验教程[M]. 北京:机械工业出版社,2021.

[31] 沈凤麟,叶中付,钱玉美. 信号统计分析与处理[M]. 合肥:中国科学技术大学出版社,2001.

[32] 孔德仁,朱蕴璞,狄长安. 工程测试与信息处理[M]. 北京:国防工业出版社,2003.

[33] 张淼. 机械工程测试技术[M]. 北京:高等教育出版社,2008.

[34] 杨将新,杨世锡. 机械工程测试技术[M]. 北京:高等教育出版社,2008.

[35] 孔德仁,朱蕴璞,狄长安. 工程测试技术[M]. 北京:科学出版社,2004.

[36] 潘宏侠. 机械工程测试技术[M]. 北京:国防工业出版社,2009.

[37] 张优云,陈花玲,张小栋,等. 现代机械测试技术[M]. 北京:科学出版社,2005.

[38] 黄长艺. 机械工程测量与试验技术[M]. 北京:机械工业出版社,2004.

[39] 刘培基,王安敏,王淑君,等. 机械工程测试技术[M]. 北京:机械工业出版社,2003.

[40] 秦树人.机械工程测试原理与技术[M]. 重庆:重庆大学出版社,2011.

[41] 胡广书. 现代信号处理教程[M]. 北京:清华大学出版社,2004.

[42] 王济. MATLAB 在振动信号处理中的应用[M]. 北京:知识产权出版社,2006.

[43] 何道清. 传感器与传感器技术[M]. 北京:科学出版社,2004.

[44] 张洪润,张亚凡. 传感器技术与应用教程[M]. 2 版. 北京:清华大学出版社,2005.

[45] 王雪文,张志勇. 传感器原理及应用[M]. 北京:北京航空航天大学出版社,2004.

[46] 郭爱芳. 传感器原理及应用[M]. 西安:西安电子科技大学出版社,2007.

[47] 高国富. 智能传感器及其应用[M]. 北京:化学工业出版社,2005.

[48] 李晓莹,张新荣,任海果. 传感器与测试技术[M]. 2 版. 北京:高等教育出版社,2004.

[49] 秦树人. 虚拟仪器[M]. 北京:中国质检出版社,2004.

[50] 陈桂明,张明照,戚红雨,等. 应用 MATLAB 建模与仿真[M]. 北京:科学出版社,2001.

[51] 柏林,王见,秦树人. 虚拟仪器及其在机械测试中的应用[M]. 北京:科学出版社,2007.